INDEX / TABLE OF CONTENTS

INTRO -------p. 5

INTELLIGENCE

    Illustration of a Derivative----------p. 9

    Coherent Brain-----------------------p. 10

    Dimensional Secrets --------------p. 11

EMOTIONS -------------------p. 13

SOCIETY-------------p. 17

POSTHUMANISM----------------------p. 23

TELALETHEIAN SCIENCE---------------p. 31

GENERAL TECHNOLOGY --------------------p. 65

Perpetual Motion ---------------p. 85

The Core System --------------p. 93

BACKGROUNDS --------------p. 142

BIO ------------------------------------ p. 186

SUBLIME ENGINEERING BY NATHAN COPPEDGE

© 1999, 2000, '01, '02, '03, '04, '05, '06, '07, '08, '09, '11, '13, '14, '15, '16, '17, '18, '19, '20, '21, 2022 Nathan Coppedge. This text compilation released 2022. Some writing has been made available elsewhere under copyright control (citation of the author is required).

SUBLIME ENGINEERING

A WORK OF COHERENT TECHNOLOGY

NATHAN COPPEDGE

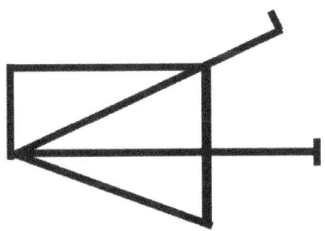

SUBLIME ENGINEERING BY NATHAN COPPEDGE

INTRODUCTION

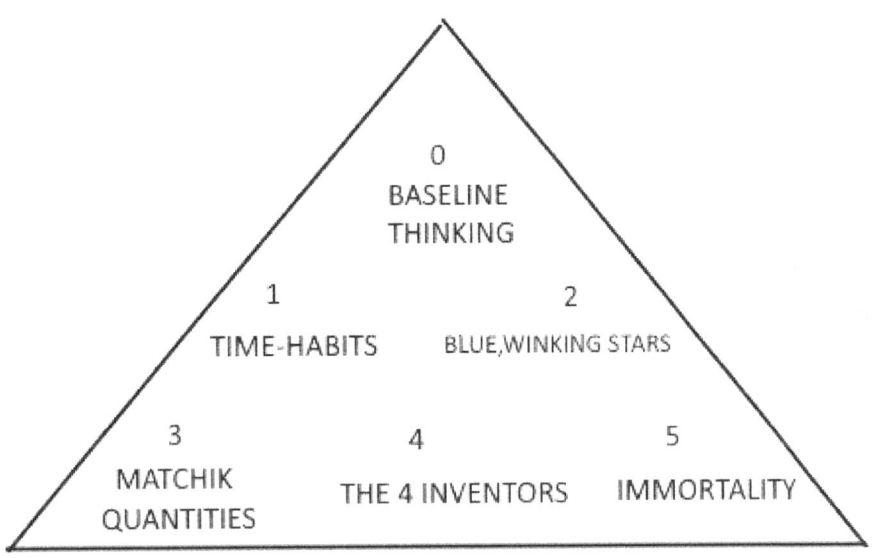

INTERPRETATION OF GOING TO THE CANDY STORE
Nathan Coppedge

///

INTELLIGENCE

SUBLIME ENGINEERING BY NATHAN COPPEDGE

SUBLIME ENGINEERING BY NATHAN COPPEDGE

DERIVATIVE

HYPER-CUBISM   Nov 2014   Nathan Coppedge

COHERENT BRAIN

Re-assessment of human ideas:

1. Physical luck. 14. Regular luck.

2. Greed for works. 5. Greed for ideas. 7. Regular greed.

3. Sufficient ideas. 6. Idea. 10. Obviated idea.

4. And 9. Mental sensations.

5. (Skip).

6. (Skip).

7. (Skip).

8. Mental-physical works.

9. (Skip).

10. (Skip).

11. And 12 Physical art.

12. (Skip).

13. Madness.

14. (Skip).

SUBLIME ENGINEERING BY NATHAN COPPEDGE
THE DIMENSIONAL SECRETS

Inventing between two earths.

Abstract humans and the earth.

Abstract inventions.

Double-inventing.

Rare inventing.

Twin inventions.

Tracing wind.

Tracing between two earths.

Balance of fire.

Tracing the traces.

Tracing the wind in the sand.

Earth wind.

Three elements.

Four elements.

Competing winds.

The Earth.

The Universe.

Four missing elements.

Missing traces of wind.

Nothing moving in wind, or land or sea.

No inspiration.

The ends of the Earth. (—December 17, 2021)

LOCAL PSYCHOLOGY DRUG CONTINUED

Mechanating.

Harrowed Form.

The Form Alien.

The Form Manisterialist.

The Form of Naga Secret.

The Form of Three Volleys.

The Form Inexorable and Form of What Is Musical.

The Form of Life and Death.

The Form of a Paradisal Glade, Enjoyment, Understandable Things.

The Form of Victory.

The Form of True Culture and Magic Mirrors.

The Living Imagination and Phenomenae Magicka.

The Essence Arcane and Reliquery Dream.

...

EMOTIONS

LOCAL PSYCHOLOGY DRUG

Insanity Wave, Clear Wave, Impossibility Wave…

- **Spell for 40 years of life: Immortality: It wasn't that it was so bad, it wasn't that it was so good, it was just different.**
- **Insanity Wave, Clear Wave, Impossibility Wave…**
- **Abstraction Reason Eternity**
- **Slicing Lemons: That Devil: Has an odd look on his face.**
- **Global Complaints, Exaggerated Norms, Collapsible Distance.**
- **Perfect standards, transcendent reality wonders, transcendent symbols. [The Functional Impossibility]**
- **Not to sop and soup: this is the basic functionality some people need.**
- **Ann Bolyn's Operant Conditioning: If someone else eats a grape I would not feel glad. —<u>Science Pills</u>**
- **[Emotional Education] It supports the development of the introspective personality but it also thrives on opportunity and feelings like intelligent development and wonders of the world. The ideal meaningful person thrives on continual emotional learning.**
- **Imagination contests: The we'd be weirdly wizards feeling.**

COHERENCE AND CHEMICAL REWARDS:

The reward for a standard logic can ideally have certain results:

- Descriptive model.
- Continuation of the logic.
- Description of a finite infinite.
- Exponentiation of some external.

These clearly do not provide much of a reward for the user's chemical system.

Let us look at the results in terms of pleasure:

- Description of pleasure: reduction to a description, requires magic.
- Continuation of pleasure: a serotonin cycle or typical process. Typical process might be enhanced but is otherwise un-exciting. Serotonin cycle comes at a cost.
- Description of a finite infinite: an exaggerated description, not authentic pleasure.
- Exponentiation of an external: enhancement of things like vision and noise. This might depend on more advanced technology before being realized except for low-stimulus events.

Now if instead we look at coherence, the rewards for a coherent logic express the following:

- Expression of a complete model.
- Containment of the model.
- Description of the absolute.
- Exponential efficiency.

Translating for pleasure:

- The sense of fulfillment (completeness). Obviously more beneficial than before.
- Containment, which is to say, continuing reward, possibly more advanced than serotonin.
- The absolute, which is to say, ultimate, or a scientific equivalent (perhaps).
- A principle of exponential efficiency, which means an enhanced effect of the above.

Coherence is a better model for chemical rewards than traditional logic.

SOCIETY

HISTORICAL QUEST SYSTEM:

You can try combining simple vs. complex plus either empirical or intellectual depending on which happened recently.

Ideally with artwork you would learn to observe nature.

With observing nature you would learn how to build clocks.

With building clocks you would learn about the mathematics of nature.

With the mathematics of nature you would learn about minimalist art.

With minimalist art you would learn about advanced science.

With advanced science you would learn about simple genius.

With simple genius you would learn about everything that exists.

With everything that exists you would learn about exceptional examples.

With exceptional examples you would learn about strange universes.

With strange universes you would learn about exemplary paths.

With exemplary paths you would learn about magical matters.

---<u>What's a good field for revolutions like Einstein's?</u> (...)

SUBLIME ENGINEERING BY NATHAN COPPEDGE

CONSUMER MODEL OF TECH

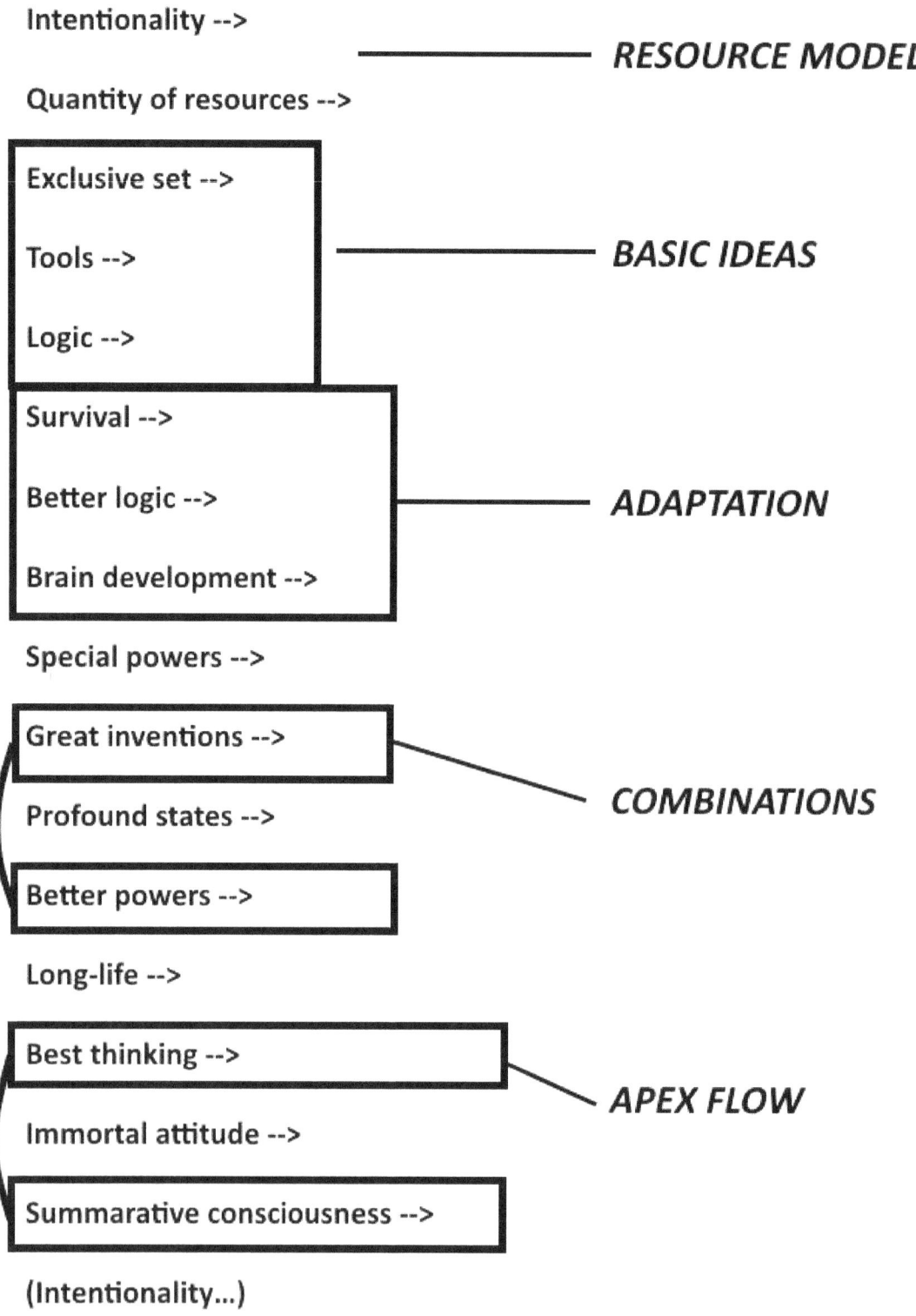

Reproducible under Nathan Coppedge

CULTURAL SUPER-STRATEGIES

Chinese:

- Martialing the arts.
- Overwhelming forces.

Phoenicians:

- Overwhelming forces.
- Religious traditions.

Egyptians:

- Religious traditions.
- Appealing culture.

Greeks:

- Appealing culture.
- Enslaving others' cultures.

Romans:

- Enslaving others' cultures.
- Consolidating power.

French:

- Consolidating power.
- Robbing the people.

British:

- Robbing the people.
- Industrialization.

Americans:

- Industrialization.
- Mass production.

Rich culture?:

- Mass production
- Perpetual motion?

ETERNAL RESOURCE MODEL:

0.0 STRATEGIC QUESTIONS

1.0 INFINITY

- Dimensions.

2.0 DESIGN

- Complexity.
- Perfection.

3.0 PLANET

- Habitable.
- Water.
- Safety.

4.0 RARITIES

- Minerals.
- Complex matter.
- Life.
- Production, Food, and Fuel

5.0 INFORMATION

- Tools.
- Communication.
- Organization.
- Analysis.
- Ideas.

6.0 CORE INVENTIONS

- Writing.
- Medicine.
- Flight.
- Electronics.
- Exponential efficiency.
- Matchik.

7.0 MODEL

- Senses.
- Symbols.
- Intelligence.
- Specialization.
- Meaning.
- Complete descriptions.
- Theory of Anything.

POSTHUMANISM

SUBLIME ENGINEERING BY NATHAN COPPEDGE

16-FOLD-PATH

0. Immortal Substance (Motto: "It's impossibly real")

1. Immortally Marvelous (Motto: "It's marvelous, so it seems impossible")

2. Eternal Immortality (Motto: "If it keeps running, it's literally a miracle")

3. Explanation of Immortality (Motto: "These perpetual motion machines take us back to infinity")

4. Diagrammatic Immortality (Motto: "We have a diagram, now it is back to the old perpetual motion machines")

5. Immortal Analysis (Motto: "We are analyzing, now it is back to the diagrams")

6. Coherent Immortality (Motto: "Back to analyzing, which will result in visual diagrams")

7. Information Immortality (Motto: "There is coherence back there somewhere")

8. Paradigmatic Immortality (Motto: "Information immortality has become archaic")

9. Content Immortality (Motto: "Now that we have content, we can focus on information")

10. Dimensional Immortality (Motto: "We have multiple dimensions, then we can create content")

11. Immortal Meaning (Motto: "We found something meaningful and meaning has dimensions")

12. Immortal Singularity (Motto: "We have thought of something singular. It might have meaning!")

13. Immortal Soothing (Motto: "There is something soothing. What singular sense emerges from it?")

14. Immortal Sublimata (Motto: "What is good about it if it does not sooth us?")

15. Immortal Subtlimata (Motto: "Now that it is realized, so subtle it seems, it might be sublime.")

SUBLIME ENGINEERING BY NATHAN COPPEDGE
THE IMPORTANCE OF COHERENCE IN THE THEORY OF ANYTHING

Impossible Humans (coherence)
    * Human coherence (fortunatemen) ----> Impossible Fortunatemen **(1)**

Coherent Fortunatemen (TOE)
    * Fortunatemen TOE (languages) ----> Coherent Languages **(2)**

TOE Language (perpetual motion)
    * Language perpetual (time) ----> TOE Time **(3)**

Perpetual motion time (infinity)
    * Timed infinities (damages) ----> Perpetual damages **(4)**

Infinite damage (rarities)
    * Damage rare (effect) ----> Infinite Effect **(5)**

Rare effect (souls)
    * Effect of the soul (advantage) ----> Rare Advantages **(6)**

Soulful advantage (emotion)
    * Advantage of emotion (madness) ----> Soulful Madness **(7)**

Emotional madness (common inventions)
    * Madness of common inventions (democracy) ----> Emotional Democracy **(8)**

Common invention democracy (polarity)
    * Democratic polarity (categorical) ----> Common Invention Categories **(9)**

Polar categories (genius)
    * Categorical genius (differences) ----> Polar Differences **(10)**

Genius difference (skill)
    * Different skills (impossibility) ----> Genius Impossibility **(1 REPEATS)**

Skills in mystery (drugs or nothing)
    * Mystery drugs (coherence) ----> Skill (Coherence) **(2 REPEATS)**

Drugs or nothing impossible (humans) ----> COHERENCE **(3 REPEATS)**

NOW FOUND TO BE A CYCLE OF 10 CATEGORIES

Reproducible under Nathan Coppedge

IMMORTALITY, TYPES OF,: The Six Immortals may be major categories of the Theory of Everything. Permuting the six categories with some irrelevant categories omitted yields:

COHERENT IMMORTALITY ATTEMPT FOR HUMANS

Soul immortality:

- Soul of immortality.
- Perpetual motion magic.
- Essence of immortality.
- Out-thinking soul.
- Party favors.

Perpetual motion immortality:

- Immortal soul and perpetual motion.
- Perpetual motion quantified.
- Immortality drug and perpetual motion.
- Immortal ideas and perpetual motion.
- Nirvana and perpetual motion.

Medical immortality:

- Sufficiency drugs.
- Secondary drugs.
- Smart drugs.
- Natural drugs.
- Prodigious drugs.

Intellectual immortality:

- Idea of immortality (helps).
- Inspired drugs (helps).
- Greatest idea of immortality (excitement).
- Transcendence (scene change, personality change).
- Infinite ideas (best of available options).

Nirvanae immortality:

- Immortal's idea (on authority and luck).
- Perpetual motion nirvana (mechanical of immortality).
- Truth be to immortality (rare inspiration).
- Highest nirvana (deservedness or divine coincidence).
- Trial-and-error nirvana (an obsessive and patient method).

Immortality of infinite variations:

- Variation on the soul (some are immortal, some not).
- Try genius mechanism (some are perpetual, in truth).
- Trial and error drugs (something will work, somehow).
- Knowledge of nirvana (divine method, difficult).
- Variations on infinity (possibility of shortcuts, divine intuition).

Perpetual motion genomics:

- Idea: Immortality is perpetual motion organics. Without perpetual motion we cannot achieve immortality. With perpetual motion math, immortality may be achievable.
- Immortality is perpetual motion organics / Perpetual motion nanorepair or organic perpetual motion nanostructures / designing a perpetual motion genome.
- Water as magical substance / Trace essential-genetics
- Some claim caffeine extends lifespan
- Time-travel / anti-dimensions / timeless modality e.g. temporal lobe function
- Stamina drugs, e.g. jiaogulan combined with ginseng and turmeric
- Higher dimensional dynamics, such as time-dynamics and cultural hyper-adaptivity, for example, immunity, privileges, beauty, tolerance, necessity ('anti-buzzard words')

Katsioulis

- Could limited hydrogen decay plus gravity explain the longevity of the universe?
- We should eat slightly heavier elements very carefully?
    - High levels of Oxygen.
    - Potassium.
    - Calcium.
    - Gold (?)
        - My brother claims it makes him stupider.
- Sleep may be a specific part of the brain which activates. The gland-based model of meta-cognition.

…

Problems, How To Solve:

- Practical problems
    - Mechanical
        - Discover all Inventions
    - Social
        - Solve philosophical problems - -> Paradox - -> Paroxysm
        - Cure pain - -> Virtuous Drugs.
        - Cure madness - -> Appropriate Minds
        - Become immortal - -> Continuous Truth

Humans are processed out with coherence, coherence may be the universe, raremen / fortunatemen may be the true souls.

'Lost in the Mental Sensor Technology' might mean enjoying immortality.

A divine human arguably would have at least 1.5 X 40 = 60 Health Points each worth 0.05 exponential or OU units of usable energy. This suggests immortals would have at least 1.5X the normal human amount of energy. ---<u>Cooking with the TOE</u>

IMMORTALITY

Jiaogulan: possibly causes eyes to eventually go black, see in dark, see negative colors when wearing sunglasses, become immortal. Jiaogulan also increases strength, stamina, and mental performance.

D + 3 = # IMMORTAL ASPECTS (SUCH AS HISTORICAL IMMORTALS OR SUBSETS)

THE REAL METHOD

For example, 1. The fortunatemen invent something. 2. There are results, 3. The results manifest , 4. The ideas are popular, 5. The power is held in common, 6. It is the power of consumption, 7.The invention becomes culture, 8. The fortunatemen is rareman.

IMMORTAL CHEMISTRY OF THE 1ST KIND

Elements - 1 = Perpetual Motion (ordinary difference + 1), this could also be seen as knowledge of elements.

Carbon - 1 = Perpetual Motion (ordinary difference + 1), this could slso be seen as knowledge of carbon, hydrogen passing upwards through carbon.

Helium + anticarbon + boron = super-perpetual, perhaps related to flying.

Elements after lithium + anticarbon + boron = exemplary advanced perpetual motion.

Antihelium + carbon + antilithium = exemplary perpetual motion perhaps related to immortal languages.

IMMORTAL CHEMISTRY of the 2nd KIND

Lithium + anticarbon + byrillium

Hydrogen + anticarbon + boron

POSTHUMAN SPELLCRAFT

BASIC NEED

Repeat "September 3, 2016" and your basic needs will be met.

BRAIN SUGAR

September 2, 1978 Mindful sugar.

EVOLUTION

Repeat "15 April, 2018"

EYES, GENERAL IMPROVEMENT OF

1. Mutter Purple fucia.

2. Master the digital. —General Improvement of Eyes (...)

HAPPINESS

Repeat "June 16, 2017" and you will find happiness.

IMMORTALITY, UNDERSTANDING

July 23, 2015. Understanding immortality and time-travel.

LUCK

Say or attract real time travelers.

MADNESS, CURE

Repeat May 27, 2019

TOOTH REPAIR, AREA OF EFFECT

Rim Telorinmo

WEIGHT LOSS

Feel underneath far right rib: reduces fat in gut

TELALETHEIAN SCIENCE

SUBLIME ENGINEERING BY NATHAN COPPEDGE

THE TELALETHEIAN SCIENCE

INTERMEDIATE CATEGORIES OF THE T.O.E. E.G. COMPLETE KNOWN TECHNOLOGIES:

OVERVIEW OF AREAS (THESE MAY OVERLAP)

August 21, 2020

Attempt to apply of the Theory of Everything and similar to chemistry using Function Spectrum and similar, and related approaches.

- **1st Unified Theory of the Theory of Everything** [A theory related to the Function Spectrum, not to be confused with the original theory, though both are seen to be valid].
- **TOE^2** (...)
- Eschmicals (from self-powered flying machines with reactive mechanisms). [concept March 5, 2021 Nathan Coppedge]

Foundational:

Macro Molecular Physics (...)

Catalog of Categories (...)

PRIMARY INDEX OF INVESTIGATION

Coherence is sublime technology (archetypal categories). [COHERENCE]

Integrity is possible magic. [EFFICIENCY]

Significance is match-ik. [DIFFERENCE]

Unification is the theory of everything. [GENERALITY]

Logic is possibility and impossibility. [COMBINATION]

The categorical is ex-nihilism. [REPRODUCTION]

Absoluteness is answers. [EXPONENTIALITY]

Standards are problems and solutions (paradoxical sets). [POLARITY]

Relevance is a paradigm (paradoxical paradox). [PLURALITY]

'Wild urges' are to attract 'villains'. [CORRELATION]

Culture is effective spell-casting. [SPECIALS]

Magical quests are contained by love. [PARADOXES]

Permanents match sublime sorcery (tokens in the weather). [DIMENSIONS]

Divine sleep is a natural allusion. [CONJUNCTIONS]

Enchanted sleep is divine captivation. [OPERATIONS]

Spell-lifes are enchanting with the soul. [2-EFF]

Wisdom is intuitive spell-casting. [2-POLARITY]

Dimensions are natures. [MULTI-PLURAL]

Efficient efficiency brings greater wisdom. [HYPER-FUNCTION]

Unfailing presence is problematic problems. [POLISH]

Nature bind is a self-solving problem. [SLIGHT]

Natural power is metaphysical semantics. [RARITIES]

Diabolical genius is a paradoxical brain. [MANIFOLDS]

Good combination is a luxury platform. [SKRIMS]

Mode of immortality is a sublime reality. [D-LEVELS]

Dynamic meaning is polar opposites. [WINDFALLS]

Meaningful meaning is the second system. [VERBS]

Coherence is absolute, all else is uncertain. [IMPRESSION]

—<u>Grand Correspondence Theories</u> (...)

SUBLIME ENGINEERING BY NATHAN COPPEDGE

PROGRAMMABLE HEURISTICS

master systems

IMPROV:
1. Better
2. Multi
3. Gen
4. Interp

EVOLVE:
1. Predict
2. Research
3. Apply

PRESERVE:
1. Mod
2. Wait
3. Codify

MIND:
1. Process
2. Stop / look
3. Engage

STEDPRGRSSiON:
1. Square 1
2. Partial metaphor.
3. Continue

MIRACLES:
1. Related to a prob.
2. It is not problem.
3. Something good.

CHEATING
1. Not standard
2. Sometimes
3. Meets reqs
4. Not bad afterall

primary systems

- COHERENCE
- EFFICIENCY · DIFFERENCE
- GENERALITY · COMBINATION · REPRODUCTION
- EXPONENTIALITY · POLARITY · PLURALITY · CORRELATION
- SPECIALS · PARADOXES · DIMENSIONS · CONJUNCTIONS · OPERATIONS
- 2-EFFICIENCY · 2-POLARITY · MULTI-PLURAL · HYPER-FUNCTION · POLISH · SLIGHT
- RARITIES · MANIFOLDS · SCRIMS · D-LEVELS · WINDFALLS · VERBS · IMPRESSION

OBJ KNOWLEDGE:
AB:CD AND AD:CB
(With POLAR OPPOSITES in DIAGONALLY OPPOSITE positions...)

PREDICTING QUESTIONS:
What QUESTION is UN-important for my POLAR OPPOSITE???

INCOHERENT KNOWLEDGE
X LOOKS OPP X...
THEN OPP X DOESN'T LOOK X.

WIZARD LOGIC:
(OBVIOUS STATEMENT, THEN) A MYSTERIOUS IRONIC THING, FOLLOWED BY AN UNEXPECTED NEGATIVE REFERENCE (TO MAGIC), FOLLOWED BY A CERTAIN CONTRADICTION.

PERFECTION:
Meaning of $(D - 1)$ in relation to the object.

DISINTEGRAL
(DEEP LANGUAGES)
KEY INSIGHT =
NEG DIFF MINUS EFF

SPECIAL VALUE THEOREM
(DEEP MATH) =
$[1 (Eff) + 0.5 (Diff)] - D$

UNIFIED EFFICIENCY =
DIFF VALUE +0.5 IF MATERIAL, MINUS 0.5 IF ABSTRACT

SOLUTION TO PARADOXES:
The POLAR OPPOSITES of every word in the BEST DEFINITION OF THE PROBLEM in the same order as the ORIGINAL WORDS.

PREDICTING ANSWERS:
The OPPOSITE SUBJECT of the question CONFIRMS the OBJECT of the question.

FORMULA FOR SOULS:
Soul of the Book: 'If you [X] qualifier [subject of X and qualifier] [opp X clarified]'
Title of Book: '[quality of X] [opp qualifier]'

THEORY OF EVERYTHING:
SET 0 > EFFICIENCY* + DIFFERENCE
*WHERE EFFICIENCY IS > 1 IF OBJECT IS ACTING... AND < 1 IF OBJECT IS BEING ACTED ON

ANTITHEORY:
ANTI-THING <=
DIFF MINUS EFF

ENERGY: $[1 / (D - 1)]$
PLUS DIFFERENCE

SYSTEMIC CONTRADICTIONS =
2 X (MIN NECESSARY DIMENSIONS FOR REALITY)

CREATIVITY:
Something related to something different finding a surprising rationale for that difference.

UNIFIED LANGUAGE FORMULA:
$1.585 \times 1.09 \wedge (D - 2)$

GENUINE PSYCHIC PREDICTIONS:
1: What usually happens may happen again.
2: Opp of 2nd concern applied to first concern.
3: Normal relation defines relation to opposites.
4: What is different has a different relation with what is different from it.
5: Opposite mode related to opposite thing.
6: Infectious meaning. If X is good, it might be good with Y.

FINELY THINGS:
DIFFERENCE PLUS MAX FUNCTION NUMBER IF > 0 ,
MIN FUNCTION NUMBER IF < 0

HIDDEN ROOMS:
ELEMENTS PLUS SETS MINUS NUMBER OF KNOWN ROOMS

FLATLAND CONJECTURE:
FOR D TO MANIFEST IN THE $(D - 2)$ WOULD CONTRADICT THE CONTINUUM BETWEEN THE $(D - 2)$ TO THE $(D - 1)$

MAJOR RESEARCH

POSSIBLY USEFUL INTERPRETATION: Minus = Lighter elements leaving gravity. Plus = Lighter elements attracted by gravity.

HUMAN EVOLUTION OPTIONS IMPROVED

(Ignoring Flying Machines)

Elements - 5 = Immortal Languages

Hydrogen focus = memory dynamic

2-Helium + anticarbon = immortal languages

Lithium + Helium + anticarbon = immortal meta-languages.

Byrillium + Helium + anticarbon = immortal knowledge languages.

Boron + Helium + anticarbon = ordinary immortal languages.

Carbon + Helium + anticarbon = perpetual motion immortal languages.

Elements - 4 = Languages.

Helium focus = Linguistics.

Hydrogen + helium + anticarbon = super-advanced immortal languages.

2-helium + anticarbon = immortal meta-languages.

Lithium + helium + anticarbon = meta-languages.

Carbon + antihydrogen + anticarbon = meta-languages.

Carbon + hydrogen + anticarbon = knowledge of languages.

Carbon + anticarbon = ordinary languages.

Elements - 3 = Knowledge.

Carbon + antilithium = knowledge.

Hydrogen + anticarbon + lithium = Linguistic knowledge.

Carbon + antihelium = linguistic knowledge.

Helium + anticarbon + lithium = Philosophy.

2-lithium + anticarbon = Ordinary knowledge.

Byrillium + anticarbon + lithium = Perpetual motion knowledge.

Elements - 2 = Ordinary Objects

Carbon - helium = ordinary objects.

Hydrogen + anticarbon + byrillium = ordinary knowledge.

Helium + anticarbon + byrillium = completely ordinary.

Lithium + anticarbon + byrillium = perpetual motion machines of the 2nd kind.

Elements - 1 = Perpetual Motion (ordinary difference + 1)

Carbon - 1 = Perpetual Motion (ordinary difference + 1)

Hydrogen + anticarbon + boron = perpetual motion of the 2nd kind.

Helium + anticarbon + boron = super-perpetual.

Elements after lithium + anticarbon + boron = exemplary advanced perpetual motion.

Antihelium + carbon + antilithium = exemplary perpetual motion.

…

HUMAN EVOLUTION OPTIONS:

Not exclusively.

Carbon + antiboron = perpetual motion immortal languages.

Carbon + antibyrillium = perpetual motion flying machine languages.

Carbon - 1 = Perpetual Motion (ordinary difference + 1)

Elements after lithium + antihydrogen = exemplary advanced perpetual motion.

…

MAIN RESEARCH:

Note: This might not make me immortal yet. Although I'm not sure why.

Elements - 5 = Immortal Languages

Helium - boron = immortal languages

Lithium - boron = immortal meta-languages.

Byrillium - boron = immortal knowledge languages.

Boron annihilation = ordinary immortal languages.

Carbon - boron = perpetual motion immortal languages.

Nitrogen - boron = flying machine immortal languages.

Element beyond nitrogen - boron = immortal super advanced flying machines languages.

Elements - 4 = Languages.

Hydrogen - byrillium = super-advanced immortal languages.

Helium - byrillium = immortal meta-languages.

Lithium - byrillium = meta-languages.

Byrillium annihilation = knowledge of languages.

Boron - byrillium = ordinary languages.

Carbon - byrillium = perpetual motion flying machine languages.

Element beyond carbon - byrillium = super-advanced flying machine languages.

Elements - 3 = Knowledge.

Hydrogen - lithium = Linguistic knowledge.

Helium - lithium = Philosophy.

Lithium annihilation = Ordinary knowledge.

Byrillium - lithium = Perpetual motion knowledge.

Boron - lithium = Knowledge of flying machines.

Element beyond boron - lithium = Knowledge of super-advanced flying machines.

Elements - 2 = Ordinary Objects

Hydrogen - helium = ordinary knowledge.

Helium annihilation = completely ordinary.

Lithium - helium = perpetual motion machines of the 2nd kind.

Byrillium - helium = ordinary flying machines.

Element beyond byrillium - helium = ordinarily advanced flying machines.

Elements - 1 = Perpetual Motion (ordinary difference + 1)

Hydrogen annihilation = perpetual motion of the 2nd kind.

Helium - hydrogen = super-perpetual.

Lithium - hydrogen = exemplary perpetual motion flying machines.

Elements after lithium - hydrogen = exemplary advanced perpetual motion.

Elements = Flying Machines Level 1 (ordinary difference + 2)

Elements as they are = flying machines.

Elements + 1 = Flying Machines Level 2 (Flying machines used as an element of a grounded perpetuum mobile).

Antibyrillium + hydrogen = immortal language of advanced flying machines.

Antilithium + hydrogen = language of advanced flying machines.

Antihelium + hydrogen = knowledge of advanced flying machines.

Antihydrogen + hydrogen = Advanced flying machines.

2-hydrogen = advanced flying machines

Elements + 2 = Flying Machines Level 3 (Flying machines flying. A large number of flying machines used as a substrate for other flying perpetuum mobile).

Anticarbon + lithium = immortal language of ultra-advanced flying machines.

Antiboron + lithium = language of ultra-advanced flying machines.

Antibyrillium + lithium = knowledge of ultra-advanced flying machines.

Lithium annihilation = ultra-advanced flying machines.

Antihelium + lithium = exemplary perpetual motion.

Antihydrogen + lithium = advanced flying machines.

Hydrogen + lithium = super advanced flying machines.

2-lithium = ultra-ultra advanced flying machines.

---Exploring Chemistry

...

SCIENCE TO COHERENCE

- Idea.
- Organize.
- Apply Theory.
- Investigate range.
- Apply Theory.
- Organize.
- Idea.

COHERENCE TO SCIENCE

- Investigate range.
- Apply Theory.
- Organize.
- Idea.
- Organize.
- Apply Theory.
- Investigate range.

—The Iteration of Science

DOMAIN-VERB THEORY

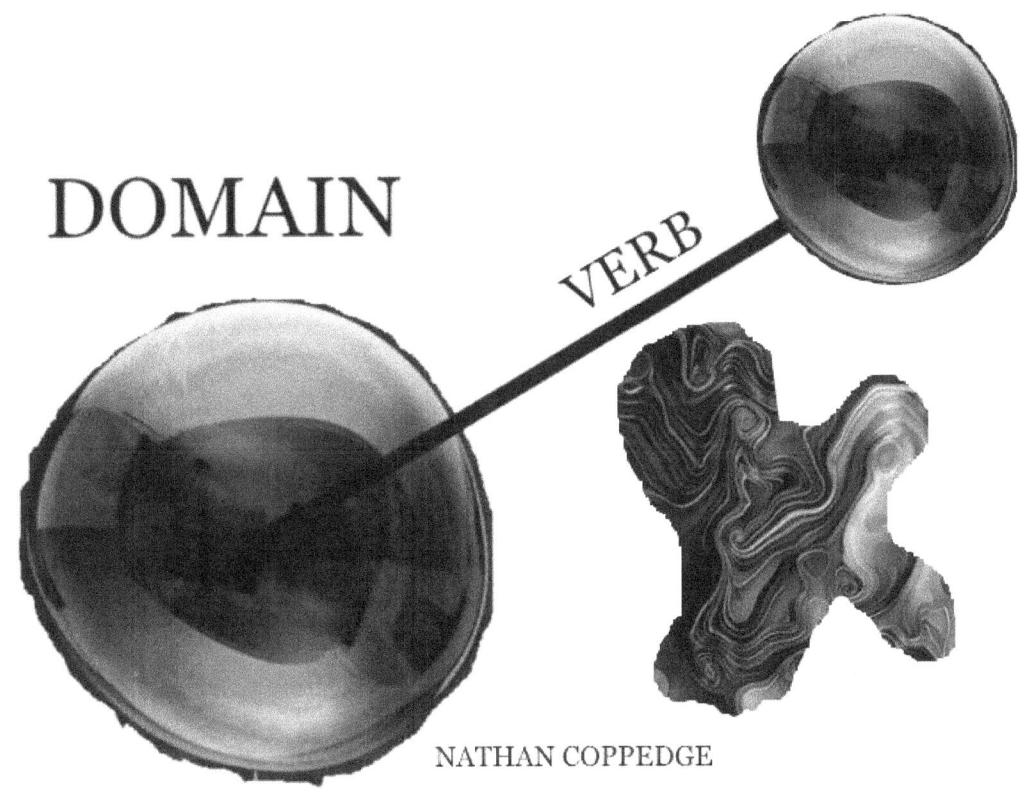

LIST OF VERBS:

Difference + (D ^ Results) - 0.5 = Estimate of Verbs [Arrived at December 22, 2020]

[Difference, Title of Difference... contents]

-5 Archaic Networks

- 1-d, 1 result = -5 + (D ^ 1) - 0.5 = -4.5 Verbs Losses
- 1-d, 2 results = -5 + (D ^ 2) - 0.5 = -4.5 Verbs Losses
- 1-d, 3 results = -5 + (D ^ 3) - 0.5 = -4.5 Verbs Losses
- 1-d, 4 results = -5 + (D ^ 4) - 0.5 = -4.5 Verbs Losses
- 2-d, 1 result = -5 + (D ^ 1) - 0.5 = -3.5 Verbs
- 2-d, 2 results = -5 + (D ^ 2) - 0.5 = -1.5 Verbs Dead-end
- 2-d, 3 results = -5 + (D ^ 3) - 0.5 = 2.5 Verbs (positive) Immortal
- 2-d, 4 results = -5 + (D ^ 4) - 0.5 = 10.5 Verbs (positive)
- 3-d, 1 result = -5 + (D ^ 1) - 0.5 = -2.5 Verbs (negative) Impossibility

- 3-d, 2 results = -5 + (D ^ 2) - 0.5 = 3.5 Verbs (positive)
- 3-d, 3 results = -5 + (D ^ 3) - 0.5 = 21.5 Verbs
- 3-d, 4 results = -5 + (D ^ 4) - 0.5 = 75.5 Verbs
- 4-d, 1 result = -5 + (D ^ 1) - 0.5 = -1.5 Verbs (negative)
- 4-d, 2 results = -5 + (D ^ 2) - 0.5 = 10.5 Verbs (positive)
- 4-d, 3 results = -5 + (D ^ 3) - 0.5 = 58.5 Verbs
- 4-d, 4 results = -5 + (D ^ 4) - 0.5 = 250.5 Verbs

-4 Draconian Networks

- 1-d, 1 result = -4 + (D ^ 1) - 0.5 = -3.5 Verbs
- 1-d, 2 results = -4 + (D ^ 2) - 0.5 = -3.5 Verbs
- 1-d, 3 results = -4 + (D ^ 3) - 0.5 = -3.5 Verbs
- 1-d, 4 results = -4 + (D ^ 4) - 0.5 = -3.5 Verbs
- 2-d, 1 result = -4 + (D ^ 1) - 0.5 = -2.5 Verbs Impossibility
- 2-d, 2 results = -4 + (D ^ 2) - 0.5 = -0.5 Verbs Abstraction
- 2-d, 3 results = -4 + (D ^ 3) - 0.5 = 3.5 Verbs
- 2-d, 4 results = -4 + (D ^ 4) - 0.5 = 11.5 Verbs
- 3-d, 1 result = -4 + (D ^ 1) - 0.5 = -1.5 Verbs (negative) Dead-end
- 3-d, 2 results = -4 + (D ^ 2) - 0.5 = 4.5 Verbs (positive)
- 3-d, 3 results = -4 + (D ^ 3) - 0.5 = 22.5 Verbs
- 3-d, 4 results = -4 + (D ^ 4) - 0.5 = 76.5 Verbs
- 4-d, 1 result = -4 + (D ^ 1) - 0.5 = -0.5 Verbs (negative) Abstraction
- 4-d, 2 results = -4 + (D ^ 2) - 0.5 = 11.5 Verbs (positive)
- 4-d, 3 results = -4 + (D ^ 3) - 0.5 = 59.5 Verbs
- 4-d, 4 results = -4 + (D ^ 4) - 0.5 = 251.5 Verbs

-3 Immortal Language

- 1-d, 1 result = -3 + (D ^ 1) - 0.5 = -2.5 Verbs Impossibility
- 1-d, 2 results = -3 + (D ^ 2) - 0.5 = -2.5 Verbs Impossibility
- 1-d, 3 results = -3 + (D ^ 3) - 0.5 = -2.5 Verbs Impossibility
- 1-d, 4 results = -3 + (D ^ 4) - 0.5 = -2.5 Verbs Impossibility
- 2-d, 1 result = -3 + (D ^ 1) - 0.5 = -1.5 Verbs Dead-end
- 2-d, 2 results = -3 + (D ^ 2) - 0.5 = +0.5 Verbs Items
- 2-d, 3 results = -3 + (D ^ 3) - 0.5 = 4.5 Verbs Generation
- 2-d, 4 results = -3 + (D ^ 4) - 0.5 = 12.5 Verbs
- 3-d, 1 result = -3 + (D ^ 1) - 0.5 = -0.5 Verbs (negative) Abstraction
- 3-d, 2 results = -3 + (D ^ 2) - 0.5 = 5.5 Verbs (positive) Eternal Time
- 3-d, 3 results = -3 + (D ^ 3) - 0.5 = 23.5 Verbs Time-Travel
- 3-d, 4 results = -3 + (D ^ 4) - 0.5 = 77.5 Verbs
- 4-d, 1 result = -3 + (D ^ 1) - 0.5 = +0.5 Verbs Items

- 4-d, 2 results = -3 + (D ^ 2) - 0.5 = 12.5 Verbs (positive)
- 4-d, 3 results = -3 + (D ^ 3) - 0.5 = 60.5 Verbs
- 4-d, 4 results = -3 + (D ^ 4) - 0.5 = 252.5 Verbs

-2 Language

- 1-d, 1 result = -2 + (D ^ 1) - 0.5 = -1.5 Verbs Dead-end
- 1-d, 2 results = -2 + (D ^ 2) - 0.5 = -1.5 Verbs Dead-end
- 1-d, 3 results = -2 + (D ^ 3) - 0.5 = -1.5 Verbs Dead-end
- 1-d, 4 results = -2 + (D ^ 4) - 0.5 = -1.5 Verbs Dead-end
- 2-d, 1 result = -2 + (D ^ 1) - 0.5 = -0.5 Verbs Abstraction
- 2-d, 2 results = -2 + (D ^ 2) - 0.5 = +1.5 Verbs Perpetual motion
- 2-d, 3 results = -2 + (D ^ 3) - 0.5 = 5.5 Verbs
- 2-d, 4 results = -2 + (D ^ 4) - 0.5 = 13.5 Verbs
- 3-d, 1 result = -2 + (D ^ 1) - 0.5 = +0.5 Verbs Items
- 3-d, 2 results = -2 + (D ^ 2) - 0.5 = 6.5 Verbs
- 3-d, 3 results = -2 + (D ^ 3) - 0.5 = 24.5 Verbs
- 3-d, 4 results = -2 + (D ^ 4) - 0.5 = 78.5 Verbs
- 4-d, 1 result = -2 + (D ^ 1) - 0.5 = 1.5 Verbs Perpetual motion
- 4-d, 2 results = -2 + (D ^ 2) - 0.5 = 13.5 Verbs
- 4-d, 3 results = -2 + (D ^ 3) - 0.5 = 61.5 Verbs
- 4-d, 4 results = -2 + (D ^ 4) - 0.5 = 253.5 Verbs

-1 Knowledge

- 1-d, 1 result = -1 + (D ^ 1) - 0.5 = -0.5 Verbs Abstraction
- 1-d, 2 results = -1 + (D ^ 2) - 0.5 = -0.5 Verbs Abstraction
- 1-d, 3 results = -1 + (D ^ 3) - 0.5 = -0.5 Verbs Abstraction
- 1-d, 4 results = -1 + (D ^ 4) - 0.5 = -0.5 Verbs Abstraction
- 2-d, 1 result = -1 + (D ^ 1) - 0.5 = 0.5 Verbs (positive) Items
- 2-d, 2 results = -1 + (D ^ 2) - 0.5 = 2.5 Verbs Immortal
- 2-d, 3 results = -1 + (D ^ 3) - 0.5 = 6.5 Verbs
- 2-d, 4 results = -1 + (D ^ 4) - 0.5 = 14.5 Verbs
- 3-d, 1 result = -1 + (D ^ 1) - 0.5 = 1.5 Verbs Perpetual motion
- 3-d, 2 results = -1 + (D ^ 2) - 0.5 = 7.5 Verbs
- 3-d, 3 results = -1 + (D ^ 3) - 0.5 = 25.5 Verbs
- 3-d, 4 results = -1 + (D ^ 4) - 0.5 = 79.5 Verbs
- 4-d, 1 result = -1 + (D ^ 1) - 0.5 = 2.5 Verbs
- 4-d, 2 results = -1 + (D ^ 2) - 0.5 = 14.5 Verbs
- 4-d, 3 results = -1 + (D ^ 3) - 0.5 = 62.5 Verbs
- 4-d, 4 results = -1 + (D ^ 4) - 0.5 = 254.5 Verbs

SUBLIME ENGINEERING BY NATHAN COPPEDGE

0 Ordinary

- 1-d, 1 result = 0 + (D ^ 1) - 0.5 = 0.5 Verbs Items
- 1-d, 2 results = 0 + (D ^ 2) - 0.5 = 0.5 Verbs Items
- 1-d, 3 results = 0 + (D ^ 3) - 0.5 = 0.5 Verbs Items
- 1-d, 4 results = 0 + (D ^ 4) - 0.5 = 0.5 Verbs Items
- 2-d, 1 result = 0 + (D ^ 1) - 0.5 = 1.5 Verbs Perpetual motion
- 2-d, 2 results = 0 + (D ^ 2) - 0.5 = 3.5 Verbs
- 2-d, 3 results = 0 + (D ^ 3) - 0.5 = 7.5 Verbs
- 2-d, 4 results = 0 + (D ^ 4) - 0.5 = 15.5 Verbs
- 3-d, 1 result = 0 + (D ^ 1) - 0.5 = 2.5 Verbs Immortal
- 3-d, 2 results = 0 + (D ^ 2) - 0.5 = 8.5 Verbs
- 3-d, 3 results = 0 + (D ^ 3) - 0.5 = 26.5 Verbs
- 3-d, 4 results = 0 + (D ^ 4) - 0.5 = 80.5 Verbs
- 4-d, 1 result = 0 + (D ^ 1) - 0.5 = 3.5 Verbs
- 4-d, 2 results = 0 + (D ^ 2) - 0.5 = 15.5 Verbs
- 4-d, 3 results = 0 + (D ^ 3) - 0.5 = 63.5 Verbs
- 4-d, 4 results = 0 + (D ^ 4) - 0.5 = 255.5 Verbs

1 Perpetual Motion

- 1-d, 1 result = 1 + (D ^ 1) - 0.5 = 1.5 Verbs Perpetual motion
- 1-d, 2 results = 1 + (D ^ 2) - 0.5 = 1.5 Verbs Perpetual motion
- 1-d, 3 results = 1 + (D ^ 3) - 0.5 = 1.5 Verbs Perpetual motion
- 1-d, 4 results = 1 + (D ^ 4) - 0.5 = 1.5 Verbs Perpetual motion
- 2-d, 1 result = 1 + (D ^ 1) - 0.5 = 2.5 Verbs Immortal
- 2-d, 2 results = 1 + (D ^ 2) - 0.5 = 4.5 Verbs Generation
- 2-d, 3 results = 1 + (D ^ 3) - 0.5 = 8.5 Verbs
- 2-d, 4 results = 1 + (D ^ 4) - 0.5 = 16.5 Verbs
- 3-d, 1 result = 1 + (D ^ 1) - 0.5 = 3.5 Verbs 1:1 machine with no mass
- 3-d, 2 results = 1 + (D ^ 2) - 0.5 = 9.5 Verbs 1:1 machine with added mass
- 3-d, 3 results = 1 + (D ^ 3) - 0.5 = 27.5 Verbs 2:1 machine
- 3-d, 4 results = 1 + (D ^ 4) - 0.5 = 81.5 Verbs 3:1 machine
- 4-d, 1 result = 1 + (D ^ 1) - 0.5 = 4.5 Verbs
- 4-d, 2 results = 1 + (D ^ 2) - 0.5 = 16.5 Verbs
- 4-d, 3 results = 1 + (D ^ 3) - 0.5 = 64.5 Verbs
- 4-d, 4 results = 1 + (D ^ 4) - 0.5 = 256.5 Verbs

SUBLIME ENGINEERING BY NATHAN COPPEDGE

2 Perpetual Motion Flying Machines

- 1-d, 1 result = 2 + (D ^ 1) - 0.5 = 2.5 Verbs Immortal
- 1-d, 2 results = 2 + (D ^ 2) - 0.5 = 2.5 Verbs Immortal
- 1-d, 3 results = 2 + (D ^ 3) - 0.5 = 2.5 Verbs Immortal
- 1-d, 4 results = 2 + (D ^ 4) - 0.5 = 2.5 Verbs Immortal
- 2-d, 1 result = 2 + (D ^ 1) - 0.5 = 3.5 Verbs
- 2-d, 2 results = 2 + (D ^ 2) - 0.5 = 5.5 Verbs
- 2-d, 3 results = 2 + (D ^ 3) - 0.5 = 9.5 Verbs
- 2-d, 4 results = 2 + (D ^ 4) - 0.5 = 17.5 Verbs
- 3-d, 1 result = 2 + (D ^ 1) - 0.5 = 4.5 Verbs Generation
- 3-d, 2 results = 2 + (D ^ 2) - 0.5 = 10.5 Verbs
- 3-d, 3 results = 2 + (D ^ 3) - 0.5 = 28.5 Verbs
- 3-d, 4 results = 2 + (D ^ 4) - 0.5 = 82.5 Verbs
- 4-d, 1 result = 2 + (D ^ 1) - 0.5 = 5.5 Verbs
- 4-d, 2 results = 2 + (D ^ 2) - 0.5 = 17.5 Verbs
- 4-d, 3 results = 2 + (D ^ 3) - 0.5 = 65.5 Verbs
- 4-d, 4 results = 2 + (D ^ 4) - 0.5 = 257.5 Verbs

3 Supported Flying Machines / 4-d Immortals

- 1-d, 1 result = 3 + (D ^ 1) - 0.5 = 3.5 Verbs
- 1-d, 2 results = 3 + (D ^ 2) - 0.5 = 3.5 Verbs
- 1-d, 3 results = 3 + (D ^ 3) - 0.5 = 3.5 Verbs
- 1-d, 4 results = 3 + (D ^ 4) - 0.5 = 3.5 Verbs
- 2-d, 1 result = 3 + (D ^ 1) - 0.5 = 4.5 Verbs Generation
- 2-d, 2 results = 3 + (D ^ 2) - 0.5 = 6.5 Verbs
- 2-d, 3 results = 3 + (D ^ 3) - 0.5 = 10.5 Verbs
- 2-d, 4 results = 3 + (D ^ 4) - 0.5 = 18.5 Verbs
- 3-d, 1 result = 3 + (D ^ 1) - 0.5 = 5.5 Verbs
- 3-d, 2 results = 3 + (D ^ 2) - 0.5 = 11.5 Verbs
- 3-d, 3 results = 3 + (D ^ 3) - 0.5 = 29.5 Verbs
- 3-d, 4 results = 3 + (D ^ 4) - 0.5 = 83.5 Verbs
- 4-d, 1 result = 3 + (D ^ 1) - 0.5 = 6.5 Verbs
- 4-d, 2 results = 3 + (D ^ 2) - 0.5 = 18.5 Verbs
- 4-d, 3 results = 3 + (D ^ 3) - 0.5 = 66.5 Verbs
- 4-d, 4 results = 3 + (D ^ 4) - 0.5 = 258.5 Verbs

4 Antiforce Mechanisms / 5-d Immortals

- 1-d, 1 result = 4 + (D ^ 1) - 0.5 = 4.5 Verbs Generation
- 1-d, 2 results = 4 + (D ^ 2) - 0.5 = 4.5 Verbs Generation
- 1-d, 3 results = 4 + (D ^ 3) - 0.5 = 4.5 Verbs Generation
- 1-d, 4 results = 4 + (D ^ 4) - 0.5 = 4.5 Verbs Generation
- 2-d, 1 result = 4 + (D ^ 1) - 0.5 = 5.5 Verbs
- 2-d, 2 results = 4 + (D ^ 2) - 0.5 = 7.5 Verbs
- 2-d, 3 results = 4 + (D ^ 3) - 0.5 = 11.5 Verbs
- 2-d, 4 results = 4 + (D ^ 4) - 0.5 = 19.5 Verbs
- 3-d, 1 result = 4 + (D ^ 1) - 0.5 = 6.5 Verbs
- 3-d, 2 results = 4 + (D ^ 2) - 0.5 = 12.5 Verbs
- 3-d, 3 results = 4 + (D ^ 3) - 0.5 = 30.5 Verbs
- 3-d, 4 results = 4 + (D ^ 4) - 0.5 = 84.5 Verbs
- 4-d, 1 result = 4 + (D ^ 1) - 0.5 = 7.5 Verbs
- 4-d, 2 results = 4 + (D ^ 2) - 0.5 = 19.5 Verbs
- 4-d, 3 results = 4 + (D ^ 3) - 0.5 = 67.5 Verbs
- 4-d, 4 results = 4 + (D ^ 4) - 0.5 = 259.5 Verbs

5 Reactive Mechanisms / 6-d Immortals

- 1-d, 1 result = 5 + (D ^ 1) - 0.5 = 5.5 Verbs
- 1-d, 2 results = 5 + (D ^ 2) - 0.5 = 5.5 Verbs
- 1-d, 3 results = 5 + (D ^ 3) - 0.5 = 5.5 Verbs
- 1-d, 4 results = 5 + (D ^ 4) - 0.5 = 5.5 Verbs
- 2-d, 1 result = 5 + (D ^ 1) - 0.5 = 6.5 Verbs
- 2-d, 2 results = 5 + (D ^ 2) - 0.5 = 8.5 Verbs
- 2-d, 3 results = 5 + (D ^ 3) - 0.5 = 12.5 Verbs
- 2-d, 4 results = 5 + (D ^ 4) - 0.5 = 20.5 Verbs
- 3-d, 1 result = 5 + (D ^ 1) - 0.5 = 7.5 Verbs
- 3-d, 2 results = 5 + (D ^ 2) - 0.5 = 13.5 Verbs
- 3-d, 3 results = 5 + (D ^ 3) - 0.5 = 31.5 Verbs
- 3-d, 4 results = 5 + (D ^ 4) - 0.5 = 85.5 Verbs
- 4-d, 1 result = 5 + (D ^ 1) - 0.5 = 8.5 Verbs
- 4-d, 2 results = 5 + (D ^ 2) - 0.5 = 20.5 Verbs
- 4-d, 3 results = 5 + (D ^ 3) - 0.5 = 68.5 Verbs
- 4-d, 4 results = 5 + (D ^ 4) - 0.5 = 260.5 Verbs

...

...

SUBLIME ENGINEERING BY NATHAN COPPEDGE

RELEVANT TO DIMENSIONAL MECHANICAL HYPOTHESIS

...

MATHEMATICAL PROPERTIES

TOE: Results >= Efficiency + Difference

Efficiency > Results - Difference

Difference > Results - Efficiency

Certainty = 0 [Proof: Certainty = TOE - Antitheory = Eff + Diff - Diff - Eff = 0].

Uncertainty = Antitheory [Proof: Antitheory + Certainty = Antitheory].

General OU = {Efficiency / [(D - 1) (Efficiency)] } + Difference

Anti-Energy = 1 - (D + Difference)

Anti-Theory <= Difference - Efficiency

Anti-Efficiency = Difference - Results

Anti-Difference = Efficiency - Results

Ideal Elements = D + 2

Ideal Principle = Negative [(D - 1) ^ 2]

Un-Ideal Elements = 2 - D

Un-Ideal Principle = [Sq rt 2 (1 - D)]

Basic Meaning = 5/32 proportion

Incomplete Meaning = 160 (constant)

Math Form = 0.10 X Absoluteness

Above Math = Qualities X 10

Languages up to 3-d = 1.585 X 1.09 ^ (D - 2 minimum zero)

Anti-Languages = [1.09 rt of (2 minus D)] / 1.585

Elements - 5 = Immortal Languages

Elements - 4 = Languages.

Elements - 3 = Knowledge.

Elements - 2 = Ordinary Objects

Elements - 1 = Perpetual Motion (ordinary difference + 1)

Elements = Flying Machines Level 1 (ordinary difference + 2)

Elements + 1 = Flying Machines Level 2 (Flying machines used as an element of a grounded perpetuum mobile).

Elements + 2 = Flying Machines Level 3 (Flying machines flying. A large number of flying machines used as a substrate for other flying perpetuum mobile).

Possible Measure of X / Missing-X in Y / Not Y

2 / Avg Speed = Observed (Theoretical)

[Sq rt of 0.5 (Time)] / Avg Speed = Detected

The first observer aims to refute. The second observer acts passively. The first particle responds quickly. The last particle is a slave.

Dimensions - Antiforces = Forces.

Dimensions - Forces = Antiforces.

Dimensions = Forces + Antiforces.

$(D^2 +1)^2$ = Grand Theory number [Former theories: (D(Elements(D+1))+1) OR ((Elements * Forces)+1)]

Subjectively God exists? Default for me is 0 [0 = false, 1 = true]

Subjectively psychopathy? Default for me is 1 [0 = false, 1 = true]

Opportunity for PMMs? [subjectively god exists + subjective psychopathy <=1, then "TRUE"]

Possibility of Tiny PMMs [If subjective psychopathy > subjectively God exists, then "MAYBE"]

Disintegral = - (Difference – Efficiency).

Special Value Theory = [1 (Efficiency) + 0.5 (Difference)] - D

Anti Disintegral = Efficiency - Difference = Antitheory

Antivalue Theory = (Dimensions+(Difference / 0.5 not a mistake) - Efficiency)

Necessary dimensions: This is constant (5) for our universe

Ideal contradictions: (10) Has been found = Necessary Dimensions X 2

Simple disintegral = 1 / Antivalue

Nodes (simpleforms) = Minus Antivalue

Verbs = [Difference+(Dimensions^Results)-0.5]

OU Formula for TOEs = [(Dimensions ^ Results) - Verbs - 1]

...

CREATING MATHEMATICAL PROPERTIES

One Approach:

<-- (-3.375) -6.75 (-3.375) <-- (-2.25) -4.5 (-2.25)<-- (=-1.5)... = -3 (Parametric dimension = -1.5) <-- (Parameter = -0.5) Dimension = -1 (Category = -0.5) <-- (Abstract = -0.5) Coherence = 0 (Matter = +0.5) --> (Unity = +0.5) Construct = 1 (Force = +0.5) --> (Force construct = +1.5) +3 (+1.5) --> (+2.25) +4.5 (+2.25) --> (+3.375) +6.75 (+3.375) -->

Parametric dimension = dimensional parameter

Parameter = dimension with abstraction

Dimension = 2 coherence without matter

Category = coherence without matter

Abstraction = negative matter

Matter = double negative coherence

Unity = coherence without abstraction

Construct = 2 coherence without abstraction

Force = construct with matter

Force construct = construct with force

CREATING THE UNIVERSAL ORGANIZATION

1. Perfection (not Determinism, not Arbitrariness, not Soullessness), 2. Insignificant event (gambits, importunity, non-linearity), 3. Mastery (trial, domination), 4. Loops (dialectics, time crystals), 5. Major attributes (character, fame, hormones, sex appeal), 6. Meaning (not Imperfection), 7. Lower attributes (Time-Travel, Telekinesis, Invisibility, Magic Wands), 8. Emotion (Persona, Intuition), 9. Core (Objective Knowledge, Perpetual Motion, World Peace), 10. Polarity (not Limitation), 11. Power (Influence, Control, Authority), 12. Consequences (Rewards, Repercussions), 13. Triviality (Leisure, Free Time), 14. Humans / turtles / retractor tools, 15. Immortality, Upkeeps (Gods, Pets, Automata), 16. Analogy (not Containment), 17. Ongoing (The Now, Stream of Consciousness, Brahma Consciousness), 18. Damage, 19. Effect (Generic Properties, Implementations), 20. Exclusivity (not Weakness), 21. Consistency (Patterns), 22. Inconsequentialism, 23. No Domination (Opportunity, Competitiveness), 24. Unsurprising (nully, boredom), 25. Complexity (not Unspannableness).

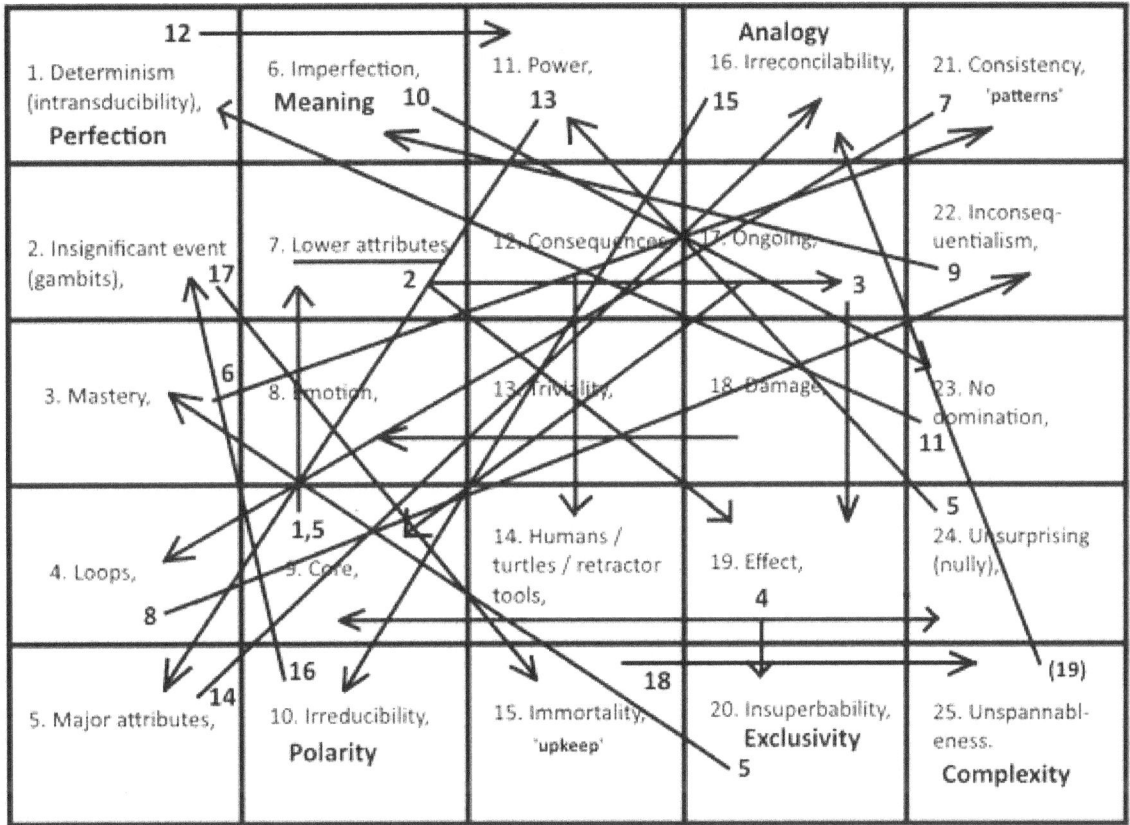

Coppedge, Nathan (2020). *The Knowledge of the Theory of Everything.*

And when difference must be added to the diagonally opposite efficiency there are really only two combinations for the whole universe:

(Universe DNA):

BCAD or DCAB AND (CDBA or ADBC) AND DACB or BACD AND (ABDC or CBDA)

OR

BCAD or DCAB AND (ABDC or CBDA) AND DACB or BACD AND (CDBA or ADBC)

…

A: Efficiency Adjective 1 (part of a polar opposite pair of adjectives).

B: Efficiency Noun 1 (part of a polar opposite pair of nouns).

C: Difference Polar opposite of Adjective 1.

D: Difference Polar opposite of Noun 1.

...

Here is the general procedure for translating knowledge into souls:

...

Knowledge: {AB:CD and AD:CB}

Basic Questions Per Query:

The question (A) is C: B-D

The question (B) is D: C-A

The question (C) is A: D-B

The question (D) is B: A-C

...

Basic Souls of Information Per Query:

The soul of A question is BCAD or DCAB.

The soul of B question is CDBA or ADBC.

The soul of C question is DACB or BACD

The soul of D question is ABDC or CBDA

CREATIING SOME TECHNOLOGY

Correct General Use of Efficiency and Difference for Determining Energy:

Number of added efficiencies equals the difference score, while added difference is always +1 or 0.

SUBLIME ENGINEERING BY NATHAN COPPEDGE

Positive:

Max: Eff = +1, Diff = +1

Min: Eff = +1, Diff = 0

Negative:

Max: Eff = -1, Diff = +1

Min: Eff = -1, Diff = 0

Note: ranges are now broadened to eliminate intermediate (0.5 level) category labels.

...

Main scores as in diagram assuming Correctly Eff +/- 1, Diff +/- 1 or 0

Reactive Mechanisms (Eff + Eff + Eff + Eff + Eff + Diff) Max 6, Min 5

Antiforce Mechanisms (Eff + Eff + Eff + Eff + Diff) Max 5, Min 4

Supported Flying Machines (Eff + Eff + Eff + Diff) Max 4, Min 3

Self-Powered Flying Machines (Eff + Eff + Diff) Max 3, Min 2

Perpetual Motion (Eff + Diff) Max 2, Min 1

Zero (Diff) Max 1, Min 0

Human Knowledge (Eff + Diff) Max 0, Min -1

Languages (Eff + Eff + Diff) Max -1, Min -2

Immortal Languages (Eff + Eff + Eff + Diff) Max -2, Min -3

Draconian Networks (Eff + Eff + Eff + Eff + Diff) Max -3, Min -4

Archaic Networks (Eff + Eff + Eff + Eff + Eff + Diff) Max -4, Min -5

ELABORATION OF TECHNOLOGIES

Main scores as in diagram assuming Correctly Eff +/- 1, Diff +/- 1 or 0

...

Reactive Mechanisms (Eff + Eff + Eff + Eff + Eff + Diff) Max 6, Min 5

- Instant Mechanisms, e.g. integrated 11, animated 10 (with reactive mechanisms, above +5 Eff).
- Faster Mechanisms, e.g. animated 10, centered 9 (with antiforce mechanisms, above +4 Eff).
- Faster Flying, e.g centered 9, unavoidable 8 (with supported flying machines, above +3 Eff).
- Fast Flying, e.g. unavoidable 8, genius 7 (with flying machines, above +2 Eff).
- Fast Movement, e.g. genius 7, clever 6 (with perpetual motion, above +1 Eff).
- Fast Transitions, e.g. clever 6, quick 5 (from zero, equivalent to above)
- Faster Processing Speed, e.g. quick 5, progress 4 (with knowledge, above - 1 Eff).
- Faster Learning, e.g. progress 4, real 3 (with languages, above - 2 Eff).
- Universals, e.g. real 3, exaggerated 2 (with immortal languages, above - 3 Eff).
- Absolutes, e.g. exaggerated 2, coherence 1 (with draconian networks, above - 4 Eff).
- Inherency / Noumena, e.g. coherence 1, certainty 0 (with arcane networks, above - 5 Eff).

Antiforce Mechanisms (Eff + Eff + Eff + Eff + Diff) Max 5, Min 4

- Antieffects (with reactive mechanisms, above +5 Eff).
- Primed effects, e.g. unavoidable +8, genius +7 (with antiforce mechanisms, above +4 Eff).
- Enhanced Volitions, e.g. genius +7, terrifying +6 (with supported flying machines, above +3 Eff).
- Exponential Volitions, e.g. terrifying +6, cleverness +5 (with flying machines, above +2 Eff).
- Root Antiforces, e.g. cleverness +5, internal movement +4 (with perpetual motion, above +1 Eff).
- Enhancements, e.g. internal movement +4, additional +3 (from zero, equivalent to above)
- Value Windfall, e.g. additional +3, windfall +2 (with knowledge, above minus 1 Eff).
- Lucky Weather, e.g. windfall +2, wild +1 (with languages, above minus 2 Eff).
- Extreme Tendency, e.g. wild +1, save 0 (with immortal languages, above minus 3 Eff).
- Recoverability / Big Save, e.g. save 0, appropriate - 1 (with draconian networks, above minus 4 Eff).
- True Dynamic, e.g. appropriate, rare (with arcane networks, above minus 5 Eff).

Supported Flying Machines (Eff + Eff + Eff + Diff) Max 4, Min 3

- Speedsters and Variants e.g. planet-pushing +9, eclipsing speed +8 (with reactive mechanisms, above +5 Eff).
- Demolitions, e.g. world-covering +8, zooming +7 (with antiforce mechanisms, above +4 Eff).
- Canopies, e.g. very quick +7, terrifying +6 (with supported flying machines, +3 Eff).
- Scarpies, e.g. very fast +6, secretive +5 (with flying machines, above +2 Eff).
- Demoniacs, e.g. clever +5, sweeping +4 (with perpetual motion, above +1 Eff).
- Impresarios, e.g. soaring +4, looming +3 (equivalent of above)
- Singularities, e.g. reliable +3, quick +2 (with knowledge, above minus 1 Eff).
- Figurations e.g. quick +2, large +1 (with language, above minus 2 Eff).
- Hulks, e.g. beastly +1, ordinary 0 (with immortal language, above minus 3 Eff).
- Visitants, e.g. regular 0, fashionable - 1 (with draconian networks, above minus 4 Eff).
- Oddmen, e.g. ornate - 1, unpredictable (with arcane networks, above minus 5 Eff).

Self-Powered Flying Machines (Eff + Eff + Diff) Max 3, Min 2

- Eschmicals, e.g. unavoidable, genius 7 (with reactive mechanisms, above +5 Eff). [concept March 5, 2021 Nathan Coppedge]
- Tumblers, e.g. genius 7, quick 6 (with antiforce mechanisms, above +4 Eff).
- Duvals, e.g. quick 6, orderly 5 (with supported flying machines, above +3 Eff).
- Wanders, e.g. disorderly 5, movement 4 (with flying machines, above +2 Eff).
- Arkites, e.g. movement 4, familiar (with perpetual motion, above +1 Eff).
- Gliders and paper airplanes, e.g. familiar, flying (with zero, equivalent to the above)
- Airplanes, e.g. flying, ethereal 1 (with knowledge, above -1 Eff)
- Mysticism, e.g. ethereal 1, important 0 (with language, above -2 Eff).
- Mythical devices, e.g. important 0, rare - 1 (with immortal languages, above -3 Eff).
- Mythical creatures, e.g. rare - 1, surprising (with diabolical networks, above -4 Eff).
- Original ideas, e.g. surprising, new (with arcane networks, above -5 Eff).

Perpetual Motion (Eff + Diff) Max 2, Min 1

- 250% Advantage, e.g. unquestionable, powerful (with reactive mechanisms, above +5 Eff).
- 200% Advantage, e.g. powerful, advantageous (with antiforce mechanisms, above +4 Eff).
- 150% Advantage, e.g. advantageous, movement (with supported flying machines, above +3 Eff).
- 100% Advantage, e.g. movement, supported (with flying machines, above +2 Eff).
- Free energy, e.g. supported, ideal (with perpetual motion, above +1 Eff).
- Energy, e.g. ideal, ratio 1 (with zero, equivalent to above)
- Exponential efficiency, e.g. ratio 1, completeness 0 (with knowledge, above - 1 Eff).
- Efficiencies, e.g. completeness 0, partial - 1 (with language, above - 2 Eff).
- Immortality, e.g. partial - 1, information (with immortal languages, above - 3 Eff).
- Special Properties, e.g. information, projection (with draconian networks, above - 4 Eff).
- Principle of Efficiency, e.g. projection, movement (with arcane networks, above - 5 Eff).

Zero (Diff) Max 1, Min 0

- Zero-point / No Energy Theory, e.g. 5-d, 4-d (with reactive mechanisms, above +5 Eff).
- Force mechanisms, e.g. 4-d, 3-d (with antiforce mechanisms, above +4 Eff).
- Perpetual motion universe, e.g. 3-d, flying (with supported flying machines, above +3 Eff).
- Levitation, e.g. flying, energy (with flying machines, above +2 Eff).
- Coasting, e.g. energy, space 0 (with perpetual motion, above +1 Eff).
- Neutral space, e.g. space 0, admission (with neutrals, equivalent to above).
- Skepticism, e.g. admission, zero -2 (with knowledge, above - 1 Eff).
- Set theory, e.g. zero -2, uncertainty -3 (with language, above - 2 Eff).
- Certainty, e.g. uncertainty -3, differences (with immortal language, above - 3 Eff).
- Limits, e.g. differentials, integrals (with draconian networks, above - 4 Eff).
- Set Limits, e.g. empty, flexible (with arcane networks, above - 5 Eff).

Human Knowledge (Eff + Diff) Max 0, Min -1

- Flexible knowledge, e.g. strong, expert (with reactive mechanisms, above +5 Eff).
- Exponential knowledge, e.g. expert, familiar (with antiforce mechanisms, above +4 Eff).
- Effective knowledge, e.g. familiar, ideal (with supported flying machines, above +3 Eff).
- Genius knowledge, e.g. ideal, exceptional (with flying machines, above +2 Eff).
- Objective knowledge, e.g. exceptional, balanced (with perpetual motion, above +1 Eff).
- Analysis, e.g. balanced, mathematical (with zero, equivalent to above)
- Exponentiality principle, e.g. mathematical, easy (with knowledge, above - 1 Eff).
- Complex languages, e.g. difficult, unfamiliar (with languages, above - 2 Eff).
- Mastering traditions, e.g. unfamiliar, old (with immortal languages, above - 3 Eff).
- Principles, e.g. new, individual (with draconian networks, above - 4 Eff).
- Wisdom, e.g. individual, humility (with arcane networks, above - 5 Eff).

Languages (Eff + Eff + Diff) Max -1, Min -2

- Movements, e.g. knowledge, familiar (with reactive mechanisms, above +5 Eff).
- Dialectics, e.g. familiar, intellectual (with antiforce mechanisms, above +4 Eff).
- Effective communication, e.g. idealistic, foolish (with supported flying machines, above +3 Eff).
- Communication, e.g. foolish, insignificant (with flying machines, above +2 Eff).
- Culture, e.g. insignificant, empty (with perpetual motion, above +1 Eff).
- Semantics, e.g. empty, emptiness (equivalent to above)
- Traditions, e.g. emptiness, unfamiliarity (with knowledge, above - 1 Eff).
- Analogies, e.g. unfamiliarity, confusion (with languages, above - 2 Eff).
- Meaning, e.g. confusion, weak body (with immortal language, above - 3 Eff).
- History, e.g. weak body, lacking strength (with draconian networks, above - 4 Eff).
- Memory, e.g. lacking strength, reliability (with arcane networks, above - 5 Eff).

Immortal Languages (Eff + Eff + Eff + Diff) Max -2, Min -3

- New ideas, e.g. greatness, way (with reactive mechanisms, above +5 Eff)
- Reactions, e.g. way, genius (with antiforce mechanisms, above +4 Eff)
- Theories, e.g. genius, creativity (with supported flying machines, above +3 Eff).
- Epiphanies, e.g. creative, possibilities (with flying machines, above +2 Eff).
- Mathematics, e.g. impossibility, tree (with perpetual motion, above +1 Eff).
- Ideas, e.g. tree, special (equivalent to above).
- Objectivity, e.g. special, generic (with knowledge, above - 1 Eff).
- Atomic facts, e.g. generic, rare (with language, above - 2 Eff).
- Universal languages, e.g. rare, susceptibility (with immortal languages, above - 3 Eff).
- Intelligence, e.g. susceptibility, dedication (with draconian networks, above - 4 Eff).
- Brilliance, e.g. dedication, mythology (with arcane networks, above - 5 Eff).

Draconian Networks (Eff + Eff + Eff + Eff + Diff) Max -3, Min -4

- Flying lab, e.g. flying, identification (with reactive mechanisms, above +5 Eff)
- Test planes, e.g. identified, useless (with antiforce mechanisms, above +4 Eff)
- Dummy aircraft, e.g. useless, expensive (with supported flying machines, above +3 Eff).
- Drones, e.g. expensive, surprise (with flying machines, above +2 Eff).
- Traps, e.g. surprise, fools (with perpetual motion, above +1 Eff).
- Life, e.g. fools, philosophy (with zero, equivalent to above)
- Inventions, e.g. philosophy, law (with knowledge, above - 1 Eff).
- Incantations, e.g. law, symbolism (with languages, above - 2 Eff).
- Enchantments, e.g. symbolism, manipulation (with immortal languages, above - 3 Eff).
- Diabolical abilities, e.g. manipulation, invisibility (with draconian networks, above - 4 Eff).
- Words of power, e.g. invisibility, destruction (with archaic networks, above - 5 Eff).

Archaic Networks (Eff + Eff + Eff + Eff + Eff + Diff) Max -4, Min -5

- Demoniacs, e.g. essentials, uniques (with reactive mechanisms, above +5 Eff).
- Sourcers, e.g. uniques, seekers (with antiforce mechanisms, above +4 Eff).
- Legendary, e.g. seekers, blind (with supported flying machines, above +3 Eff).
- Well-to-do, e.g. blind, unfamiliar (with flying machines, above +2 Eff).
- Resourcefuls, e.g. unfamiliar, lost (with perpetual motion, above +1 Eff).
- History, e.g. lost, helpless (with zero, equivalent to above).
- Great apes / great thinkers, e.g. helpless, missing hex (with knowledge, above - 1 Eff).
- Hypotheses, e.g. missing hex, slavery (with languages, above - 2 Eff).
- Schematic reality, e.g. missing hex, missing world (with immortal languages, above - 3 Eff).
- Substantial basis, e.g. missing world, missing center (with draconian networks, above - 4 Eff).

—Applied Function Spectrum

TECHNOLOGY

SUBLIME ENGINEERING BY NATHAN COPPEDGE

METHODS FOR BLACK SWANS

THE GENERAL SECRETS OF IDEAS ARE DESCRIBED BY:

[PREMIER INTELLECTUAL DIALECTIC]

1. Take a good example, and create several versions of it.

Example:

My primary treatment is to translate or expand Ockham to include philosophical razors designed to provide standards for how to do philosophy: Book of Razors

2. With your experience, extract a higher principle from the original example.

Example:

One perspective perhaps related to this is that Ockham can also have a 'higher translation' in terms of logical or mechanical (etc.) efficiency.

3. Greatly improve the higher principle by adding another factor, for example, simply doubling it.

Example:

We can then use the principle of efficiency ingeniously to arrive at exponential efficiency.

4. Now, use the improved higher standard as a platform for a body of very new ideas.

Example:

Exponential efficiency can then be used as a platform concept for masterful fulfillment of the logical and mechanical criteria.

5. Find the best general categories within the new system / platform.

Example:

This leads to the general concepts of preferred knowledge and continuous motion machines,

6. Now translate the general categories using your understanding of the general and specific meaning.

Example:

...equals objective knowledge and perpetual motion.

IDEAL INVENTIONS / GENERAL CONCEPTS

Is it related to philosophy, inventing, poetry, or art?
ART CONCEPT = _____
[For example, Cubism]

Does it reform reality? [Property of previous] =
_____
[For example, Hyper-dimensions]

Form a Neologism = _____
[For example, Hyper-Cubism]

Change to seem 'Classic Nathan' _____
[For example, The Metaphysical Art]

GREATNESS PAPER

_____

WEALTH (0 - 1) = _____
(0 STARVING, 0.25 UNSKILLED, 0.5 JOBS, 0.75 RICH, 1 TRILLIONAIRE MATERIAL)

FAME (0 - 1) = _____
(0 OUTCAST, 0.25 UNPOPULAR, 0.5 POPULAR WITH FRIENDS, 0.75 FAMOUS, 1 GREAT)

ORIGINALITY (0 - 1) = _____
(0 CAN'T DRAW, 0.25 TOO NORMAL, 0.5 SLIGHTLY CREATIVE, 0.75 PROLIFIC, 1 GENIUS)

TOTAL SCORE (MAX 2.5) = _____ / 3

TOTAL > 1 EFFEMINATE, UGLY QUALITY, FEMALE, OR BAD

TOTAL > 1.5 CRIMINAL, ABUSED, NO SEX LIFE, OR DISASTER STRIKES

TOTAL > 2 ORIGINAL CHEAT _____
TAKEN: STUPID GENIUS: NATHAN COPPEDGE,
*   RICH LIAR: HENRY FORD*
*   TORTURED FAME: JESUS*

HISTORY OF IDEAS PAPER
START ANYWHERE, ARRANGE CHRONOLOGICALLY
These refer to rough dates of each invention as a science.

Technological Complex is	Technological Complex is
Technological Simple is	Technological Simple is
Artistic Simple is	Artistic Simple is
Artistic Complex is	Artistic Complex is
Cosmological Complex is	Cosmological Complex is
Cosmological Simple is	Cosmological Simple is
Physical Simple is	Physical Simple is
Physical Complex is	Physical Complex is
A New Concept is	A New Concept is
Technological Complex is	Technological Complex is
Technological Simple is	Technological Simple is
Artistic Simple is	Artistic Simple is
Artistic Complex is	Artistic Complex is
Cosmological Complex is	Cosmological Complex is
Cosmological Simple is	Cosmological Simple is
Physical Simple is	Physical Simple is
Physical Complex is	Physical Complex is
A New Concept is	A New Concept is

THERE ARE COMPLEX RESULTS FOR WHAT IS CALLED THE PINNACLE THEORY MODEL:

T.O.E.

- KNOWLEDGE: Results (1,2,3…) = Eff + Difference
- PERPETUAL MOTION: Results = Eff (1,2,3…) + Difference
- FUNCTION SPECTRUM: Results = Eff + Difference (1,2,3…)
- Unified—!

Anti-Theory:

- Anti-Thing (1,2,3…) <= Difference - Efficiency
- Anti-Thing <= Difference (1,2,3…) - Efficiency
- Anti-Thing <= Difference - Efficiency (1,2,3…)

Efficiency:

- Efficiency (1,2,3…) >= Results – Difference
- Efficiency >= Results (1,2,3…) – Difference
- Efficiency >= Results – Difference (1,2,3…)

Anti-Efficiency:

- Anti-Efficiency (1,2,3…) <= Difference - Results
- Anti-Efficiency <= Difference (1,2,3…) - Results
- Anti-Efficiency <= Difference - Results (1,2,3…)

Difference:

- Difference (1,2,3…) >= Results – Efficiency
- Difference >= Results (1,2,3…) – Efficiency
- Difference >= Results – Efficiency (1,2,3…)

Anti-Difference:

- Anti-Difference (1,2,3…) <= Efficiency - Results
- Anti-Difference <= Efficiency (1,2,3…) - Results
- Anti-Difference <= Efficiency - Results (1,2,3…)

Forces:

- # Forces (1,2,3...) = # Dimensions - # Antiforces
- # Forces = # Dimensions (1,2,3...) - # Antiforces
- # Forces = # Dimensions - # Antiforces (1,2,3...)

Antiforces:

- # Antiforces (1,2,3...) = # Dimensions - # Forces
- # Antiforces = # Dimensions (1,2,3...) - # Forces
- # Antiforces = # Dimensions - # Forces (1,2,3...)

Dimensions:

- # Dimensions (1,2,3...) = # Forces + # Antiforces
- # Dimensions = # Forces (1,2,3...) + # Antiforces
- # Dimensions = # Forces + # Antiforces (1,2,3...)

Anti-Dimensions:

- # Anti-Dimensions (1,2,3...) = # Antiforces - # Forces
- # Anti-Dimensions = # Antiforces - # Forces (1,2,3...) +
- # Anti-Dimensions = # Antiforces (1,2,3...) - # Forces

Disintegral:

- Disintegral (1,2,3...) = - (Difference – Efficiency)
- Disintegral = - (Difference (1,2,3...) – Efficiency)
- Disintegral = - (Difference – Efficiency (1,2,3...))

Anti-Disintegral or Abstract Efficiency:

- WAR: Results (1,2,3...) = - Difference + Efficiency
- GENERAL AND SPECIAL TRANSLATION: Results = - Diff (1,2,3...) + Efficiency
- EFFICIENCY SPECTRUM: Results = -Diff + Efficiency (1,2,3...)
- Disunified—!

Super-Disintegral:

- Super-Disintegral (1,2,3...) = - (Inf Diff - Inf Eff)
- Super-Disintegral = - (Inf Diff (1,2,3...) - Inf Eff)
- Super-Disintegral = - (Inf Diff - Inf Eff (1,2,3...))

Anti-Super-Disintegral:

- Anti-Super-Disintegral (1,2,3...) = - (Inf Eff - Inf Diff)
- Anti-Super-Disintegral = - (Inf Eff (1,2,3...) - Inf Diff)
- Anti-Super-Disintegral = - (Inf Eff - Inf Diff (1,2,3...))

Min Results:

- Min Results (1,2,3...) = (Max Eff / 2) + Diff
- Min Results = (Max Eff (1,2,3...) / 2) + Diff
- Min Results = (Max Eff / 2) + Diff (1,2,3...)

Max Results:

- Max Results (1,2,3...) = Min Eff + Diff
- Max Results = Min Eff (1,2,3...) + Diff
- Max Results = Min Eff + Diff (1,2,3...)

Min Efficiency:

- Min Eff (1,2,3...) = Results - Diff
- Min Eff = Results (1,2,3...) - Diff
- Min Eff = Results - Diff (1,2,3...)

Max Efficiency:

- Max Eff (1,2,3...) = (Min Results - Diff) X 2
- Max Eff = (Min Results (1,2,3...) - Diff) X 2
- Max Eff = (Min Results - Diff (1,2,3...)) X 2

Flying Max Results

- Flying Max Results (1,2,3...) = (Min Eff) + 2 Eff - 1
- Flying Max Results = (Min Eff (1,2,3...)) + 2 Eff - 1
- Flying Max Results = (Min Eff) + 2 Eff (1,2,3...) - 1

Flying Min Results

- Flying Min Results (1,2,3...) = (Max Eff / 2) + 2 Eff - 1
- Flying Min Results = (Max Eff (1,2,3...) / 2) + 2 Eff - 1
- Flying Min Results = (Max Eff / 2) + 2 Eff (1,2,3...) - 1

THE THEORY OF EVERYTHING RESULTED IN A CONDENSED SET OF ATTRIBUTES WHICH ALSO SERVES AS A THEORY OF ENERGY:

- Math + TOE.
- Wish + Perpetual Motion.
- TOE + Elements.
- Perpetual motion + Meaning.
- Elements + Function.
- Meaning + Energy.
- Function + Variation.
- Energy + Language.
- Variation + Psychic.
- Language + Organization.
- Psychic + Species.
- Organization + Set.
- Species + Resources.
- Set + Sufficiency.
- Resources + Math.
- Sufficiency + Wish.

APPLIED GENERALIZATION

SUBLIME ENGINEERING BY NATHAN COPPEDGE

X UNIVERSE 0 D = 11 #13 IMPOSSIBILITY RESULTS= INF, EFF= NEG INF, DIFF= IMPOSSIBLE	UNIVERSE 5 D = 6 X #18 RARITIES RESULTS= FIN, EFF= NEG INF, DIFF = INF	UNIVERSE10 D = 4 #23 ABSTRACTION RESULTS= 0, EFF= NEG INF, DIFF = INF	UNIVERSE 15 D = 2 X #3 LANGUAGE RESULTS= NEG FIN, EFF= NEG INF. DIFF = INF	UNIVERSE 20 D = NEG1 #8 INSANITY RESULTS= NEG INF, EFF= RESULTS DIFF= 0
UNIVERSE 1 D = 10 #14 COHERENCE RESULTS= INF, EFF= NEG FIN DIFF= RESULTS	UNIVERSE 6: D = 0 #19 INFORMATION RESULTS = FIN EFF= NEG FIN DIFF = -(EFF) + RESULTS	UNIVERSE 11: D = 0 #24 MAINTENANCE RESULTS = ZERO EFF= NEG FIN DIFF = FIN	UNIVERSE 16: D = 0 #4 TIMED EFFECTS RESULTS = NEG FIN EFF = NEG FIN DIFF = -(EFF)+RESULTS	UNIVERSE 21: D= NEG2 #9 FRINGES RESULTS= NEG INF, EFF= NEG FIN. DIFF = RESULTS
UNIVERSE 2 D = 9 #15 THEORIES RESULTS=INF, EFF= 0, DIFF= RESULTS	UNIVERSE 7: D = 0 #20 ACTIVITY RESULTS = FIN EFF = 0 DIFF = RESULTS	UNIVERSE 12: D = 0 #25 NEUTRALS RESULTS = ZERO EFF = ZERO DIFF = ZERO	UNIVERSE 17: D = 0 #5 DAMAGE RESULTS = NEG FIN EFF = 0, DIFF = RESULTS	UNIVERSE 22 : D=NEG3 #10 CATEGORIES RESULTS= NEG INF, EFF= 0, DIFF = RESULTS
UNIVERSE 3 D = 8 #16 STANDING WAVES RESULTS= INF EFF= FIN, DIFF= RESULTS	UNIVERSE 8: D = 0 #21 CORES RESULTS = FIN EFF = FIN DIFF = RESULTS - EFF	UNIVERSE #1 COMPLEX BODIES RESULTS = AVG ZERO EFF = FIN DIFF = AVG 0 - EFF	UNIVERSE 18: D = 0 #6 EFFECTS RESULTS = NEG FIN EFF = FIN DIFF= NEGEFF+ RESULTS	UNIVERSE 23: D = NEG4 #11 LUXURIES RESULT = NEG INF, EFF = FIN, DIFF= RESULTS
UNIVERSE 4 D=7 #17 DISINTEGRALS RESULTS= INF, EFF= RESULTS DIFF= 0	UNIVERSE 9 D = 5 X #22 LIMITS RESULTS= FIN, EFF= INF, DIFF= NEG INF	UNIVERSE 14 D = 3 #2 IMMORTALITY RESULTS 0, EFF= INF, DIFF= NEG INF	UNIVERSE 19 D = 1 #7 ADVANTAGES RESULTS= NEG FIN, EFF= INF, DIFF= NEG INF	X UNIVERSE 24 D = NEG5 #12 REALITIES RESULTS= NEG INF, EFF= INF, DIFF= IMPOSSIBLE

Free and must remain non-proprietary, Nathan Larkin Coppedge

THE FIRST CATEGORY IS IMPOSSIBILITY:

5/32

IMPOSSIBILITY FORMULAS

ANTI-THING <=
DIFFERENCE - EFFICIENCY

GENERAL IMPROBABILITY

$- \mid 2 / D - (\text{results} / (OU + ((D \wedge \text{Results}) - (D + 5 - 1)))) \mid$

$D = 11$

THE SECOND CATEGORY IS COHERENCE:

15/17 D = 10
COHERENCE EQUATIONS

**GIVEN QUESTION (A) is C: BD
THEN BCAD and / or DCAB**

**GIVEN QUESTION (B) is D: CA
THEN CDBA and / or ADBC**

**GIVEN QUESTION (C) is A: DB
THEN DACB and / or BACD**

**GIVEN QUESTION (D) is B: AC
THEN ABDC and / or CBDA**

THE THIRD CATEGORY IS THEOIRES:

25/32   D = 9

THE THEORY OF ANYTHING

**TOTAL RESULTS
>= TOTAL EFFICIENCY *
+ TOTAL DIFFERENCE**

* WHERE DIFFERENCE = RESULTS - EFFICIENCY,
AND WHERE EFFICIENCY SUMS TO < 1
IF TOPIC IS ACTED ON,
AND SUMS TO > 1 IF TOPIC IS ACTING

THE FOURTH CATEGORY IS STANDING WAVES:

225/17
PERPETUAL MOTION MACHINES

MIN HEAVIER MASS = (MAX LVG / 2) + 1

MAX HEAVIER MASS = MIN LVG + 1

MIN LVG = MAX HEAVIER MASS - 1

MAX LVG = (MIN HEAVIER MASS -1) X2

OVER-UNITY = HEAVIER MASS RNG / LVG RATIO + 1 X 100 (%)

SMALLER MASS = 1X

D = 8

FIFTH CATEGORY: THE OVERALL ATTRIBUTES TRANSLATE:

625/32   D = 7

UNIVERSE 4: MAJOR ATTRIBUTES

TOE 1. MATH 2. ELEMENTS	**FUNCTION** 1. ELEMENTS 2. VARIATION	**PSYCHIC** 1. VARIATION 2. SPECIES	**RESOURCES** 1. SPECIES 2. MATH
PERPETUAL MOTION 1. WISH 2. MEANING	**ENERGY** 1. MEANING 2. LANGUAGE	**ORGANIZ-ATION** 1. LANGUAGE 2. SET	**SUFFICIENCY** 1. SET 2. WISH
ELEMENTS 1. T.O.E. 2. FUNCTION	**VARIATION** 1. FUNCTION 2. PSYCHIC	**SPECIES** 1. PSYCHIC 2. RESOURCES	**MATH** 1. RESOURCES 2. TOE
MEANING 1. PERPETUAL MOTION 2. ENERGY	**LANGUAGE** 1. ENERGY 2. ORGANIZ-ATION	**SET** 1. ORGANIZ-ATION 2. SUFFICIENCY	**WISH** 1. SUFFICIENCY 2. PERPETUAL MOTION

THE SIXTH CATEGORY TRANSLATES AS RARIFIED PROPERTIES:

25/32

FORMULAS FOR RARITIES

DRAGON'S TREASURE LOGIC

SOMETHING INTELLECTUAL YOU LOVE THE MOST, FOLLOWED BY A SYMBOL OF YOUR POWER

FERTILE GROUND

WHAT YOU LOVE THE MOST, FOLLOWED BY A CHEAP VERSION OF THAT.

$$D = 6$$

THE SEVENTH CATEGORY TRANSLATES AS INFORMATION

15/17   D = 0

SOUL FORMULAS

SOULS OF NAMES

TITLE OF BOOK = '[QUALITY OF X] [OPP QUALIFIER]'

SOUL OF BOOK = 'IF YOU [X] QUALIIFIER [SUBJECT OF X AND QUALIFIER] [OPP X CLARIFIED]

INTELLECTUAL SOUL

THE REMAINING PROBLEM IN THE SUBJECT'S ATTEMPT TO FIND COHERENCE

THE EIGHTH CATEGORY TRANSLATES AS ACTIVITY:

5/32   D = 0
EMOTIONAL FORMULAS

1. FUNDAMENTAL MISUNDERSTANDING (CARING TOO MUCH FOR INERT THINGS, EXAGGERATING SIMILARITY BETWEEN DIFFERENT THINGS.

2. MOTHERING INSTINCT APPLIES TO THE WHOLE COSMOS. FOR EXAMPLE, "IT WANTS TO BE ON EARTH AGAIN".

3. "SOMETIMES I WANT TO BE LIKE IT." FANTASTIC LIE, MADDENING REVERSAL.

4. EXTREME SENTIMENTALITY. WHY ME? COULDN'T IT BE BETTER? I DO REALIZE! WHY CAN'T WE BE THE SAME? DOESN'T HE FEEL LIKE ME? WHEN WILL THINGS CHANGE? DOESN'T HE KNOW? WHAT IS EVER THE SAME? WILL I FEEL BETTER?

5. STUPID MAN. CRY LIKE A LITTLE GIRL.

THE NINTH CATEGORY TRANSLATES AS BASIC INVENTIONS:

625/32   D = 0

CORE INVENTIONS

1. SOMEONE CONTRADICTS YOUR INFORMATION

2. CALL UP INFORMATION

3. THE WAY YOU CALL UP INFORMATION ANTICIPATES THE NEW INVENTION

THE TENTH CATEGORY TRANSLATES AS LIMITS

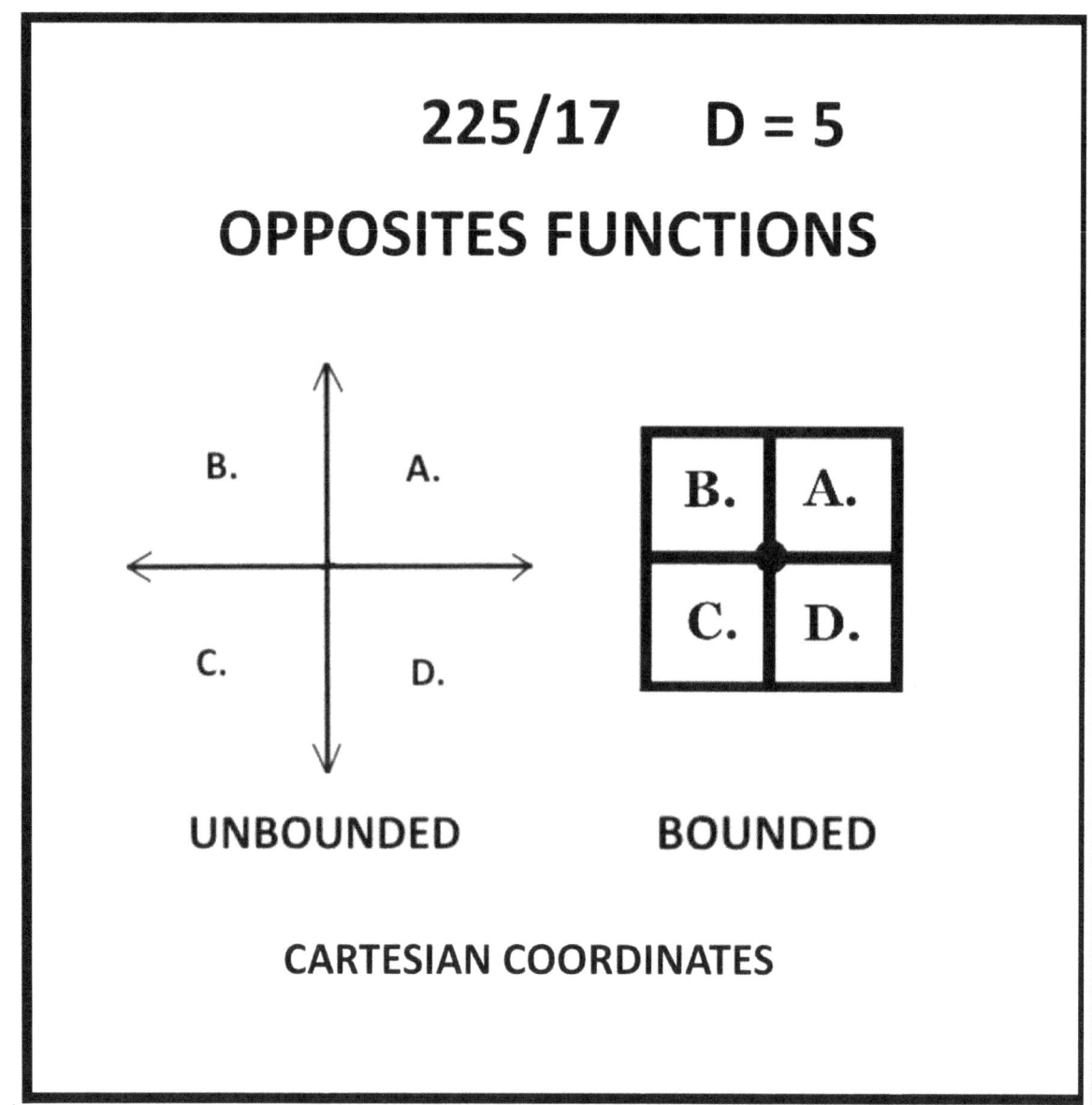

THE ELEVENTH CATEGORY TRANSLATES AS ENERGY OR INSPIRATION:

5/32   D = 4

ENERGY / BRILLIANT IDEAS

WHAT IS OBVIOUS? [INPUT]

WHAT IS TRIVIAL IN THIS TIME? [INPUT]

WHAT IS THE BETTER 2-STEP OF (TRIVIAL)?

WISE ANSWER [INPUT]

WHAT IS MOST REQUIRED FOR TRIVIAL???

YOU WILL FIND IT IS [WISE ANSWER]

PRIMARY INVENTION: [WISE ANSWER]

THAT WISHES FOR [TRIVIAL]

THE TWELFTH CATEGORY TRANSLATES AS MAINTENANCE BUT IT TURNS OUT IS UNNECESSARY

SUBLIME ENGINEERING BY NATHAN COPPEDGE

PERPETUAL MOTION HAS BEEN OBSERVED TO HAVE A VARIETY OF FORMS THUS FAR…

CONVENTIONAL AND UNCONVENTIONAL VERTICAL WHEELS

- BHASKARA
- CLAIM INVOLVING SPECIAL RATCHETS, ETC, E.G. T-BAR IDEA.
- ARACHNID ANALOG BY B COPPEDGE AND K RUBEN.
- SEE ALSO WATER DEVICES.

CONVENTIONAL CHAINS AND BUOYS

- UNBALANCED BUOYS USING PRESSURE FROM SIDE OF SPONGE.
- SO-CALLED UNBALANCED CHAIN
    - SPIRAL CHAIN, SUCH AS CURVED RAIL DEVICE, THERE IS AN EXAMPLE OF THIS WHICH MAY HAVE <113% OU.
    - CHAIN FEEDER DEVICES, SOME OF THESE ARE INTRIGUING.

WATER DEVICES

- MODIFICATIONS OF FRANK TATAY, E.G. WIDE LOWER TANK ABSORBS A TOWER OF PRESSURE.
- INGENIOUS WATER WHEEL TIPS WATER OVER TOP THEN APPLIES LEVERAGE, BY K. RUBEN.
- USE OF REPEATED INTERNAL BUOYANCY TO LIFT LEVER RODS UPWARDS AND KEEP THEM SIDEWAYS.
- SPONGE DEVICES
    - SPONGE WHICH DIPS SOAKING CORD WHEN WATER EVAPORATES FROM SPONGE.
    - SPONGE WHICH ABSORBS ON A SLANT, FEEDING A SMALL STREAM

- JEN'S BUBBLEVATOR, USE OF BUOYANCY ON LAND TO CREATE LEVERAGE WHICH IS THEN SHIFTED AGAIN AFTER LEVERAGE IS APPLIED.

- UNDERWATER ESCHER MACHINE: USE OF A BUOY IN PLACE OF A WEIGHT, USED UPSIDE-DOWN IN A SPECIAL PROPORTIONAL ANGLE UNDER CERTAIN CONDITIONS.

- MODULAR BUOY DEVICE, USE OF BUOYS UPSIDE DOWN IN PLACE OF WEIGHTS IN A SIMILAR ARRANGEMENT TO A MODULAR PERPETUAL MOTION.

- UNDERWATER BUBBLEVATOR: A BUBBLEVATOR USING A MASS IN PLACE OF A BUBBLE TO CREATE THE SAME EFFECT AS THE BUBBLEVATOR, EXCEPT UPSIDE-DOWN, UNDERWATER, WITH BUOYS ATTACHED TO THE ENDS OF THE LEVERS.

- USE OF UNDERWATER PERPETUAL SUBMARINES TO CREATE UNDERWATER ELEVATORS.

SPIRAL DEVICES

- DIFFERENTIAL PENDULUM SPIRAL WHEEL.

- SPIRAL CONE DEVICES.

- SPIRAL MAGNET USING HORIZONTAL PRESSURE TO DIRECT BALL UPWARDS.

HORIZONTAL WHEELS

- THE TILT MOTOR, A HORIZONTAL WHEEL OPERATED USING LEVERS WHICH CREATE A SLANT WHICH PURSUES THE ROLLING CONE AT 90 DEGREES.

- SLANTED PULLEY DEVICE, A DEVICE OPERATED BY AN UNBALANCED MARBLE WHICH IS LIFTED USING A WEIGHTED PULLEY WHILE SUPPORTED, BUT HAS AN ADVANTAGE WHEN APPLYING FULL WEIGHT TO THE SLANTED WHEEL.

- A DESIGN REVEALED AT SCHOOL WHICH SUGGESTS 3 SPOKES CONSTANT APPLICATION COULD HAVE AN ADVANTAGE ON 5 SUPPORTED WITH ALL HAVING WHEELS AND INTERNAL BLOCKING AT A T-JUNCTION, AND THE WHEEL BEING POSITIONED ON A SLANT IN A PARTICULAR WAY TO ALLOW THE 3 ARMS TO DROP WITH FULL PRESSURE, AND THE 5 ARMS TO BE LIFTED ALONG A TRACK SLIGHTLY ABOVE THEIR BLOCKING POINTS.

VERTICAL LEVERS

- THE VERTICAL LEVER DEVICE: A DEVICE USING A COUNTERWEIGHT POSITIONED BELOW WITH THE LONG END DIRECTED UPWARDS IN ABOUT A 4:1 RATIO, ACTING ON A 1X SPHERE WITH 3X – 5X COUNTERWEIGHT OPTIMIZED AT 4X COUNTERWEIGHT APPROXIMATELY, IN A VERY SPECIFIC RATIO MENTIONED IN A DIAGRAM FOR THE IMPROVED VERTICAL LEVER. THE COUNTERWEIGHT SHOULD BE POSITIONED STRAIGHT OUT FROM THE END, NOT AT A T-JUNCTURE, HOWEVER THE WAY THE COUNTERWEIGHT IS ATTACHED CAN AFFECT HOW IT WORKS. SUPPORT AND NON-SUPPORT IS USED AS USUAL FOR THE TWO DIRECTIONS OF MOVEMENT OF THE SPHERE.

- THE NIBW4: THIS IS THE SAME AS THE VERTICAL LEVER EXCEPT WITH THE COUNTERWEIGHT ATTACHED ABOVE AND THE LONGER END RUNNING DOWNWARDS WITH A TRACK BUILT PROPERLY BELOW.

HORIZONTAL LEVERS

- REPEAT LEVER 2 WAS AN EARLY DESIGN AIMING TO MAKE USE OF A TRIANGULAR TRACK TO DIRECT A SMALL BALL TO DIFFERENT POSITIONS OF LEVERAGE

- THE SWIVEL LEVER DEVICE: THIS DEVICE USES A SLANTED LEVER WHICH SWAYS SIDEWAYS USING A COUNTERWEIGHT, WITH THE LONG END PUSHING A BALL UP A VERY SLIGHT INCLINE, AND THEN HAVIG THE BALL DEFLECT INWARDS SO THAT IT CAN APPLY PRESSURE TO A BASKET ON THE LEVER, AND THEREBY LIFT THE COUNTERWEIGHT. IT HAS BEEN SHOWN TO MOSTLY WORK, THOUGH IT DEPENDS ON MANY DETAILS.

- CRESCENT LEVER: AN INGENIOUS DESIGN MAKES USE OF A PROTRACTOR-SHAPED MOBILE ELEMENT ATTACHED TO A LEVER. IT IS THOUGHT IF THE PROPORTIONS OF WEIGHT ARE CORRECT, WITH A HEAVY COUNTERWEIGHT AT SHORT DISTANCE ON THE OPPOSITE END, A SMALL BALL CAN BE GUIDED BY THE SLANT OF THE PROTRACTOR-SHAPED ELEMENT AND THEN WHEN IT REACHES A STRAIGHT PORTION THAT IS MORE STEEPLY SLOPED, IT CAN SLIP ALONG THE ANGLE OF THE PROTRACTOR SHAPE AND RETURN TO THE BEGINNING, CREATING AS USUAL AN ENDLESS CYCLE.

MODULAR PMMS

- THE 'MMM' OR MOTIVE MASS MACHINE IS INSPIRED BY DOMINOES WHICH CAUSE THEMSELVES TO RESET. A 'DIFFERENCE WEIGHT' APPLIES PRESSURE TO A SHALLOW BALANCE, WHICH IS THEN USED TO APPLY MOSTLY HORIZONTAL PRESSURE TO THE NEXT DIFFERENCE WEIGHT. IF THE DIFFERENCE WEIGHTS MOVE IN A NARROW CENTRAL SECTION OF THE SHALLOW BALANCE, THEN THE FORCE APPLIED MAY EXCEED WHAT IS NECESSARY TO SHIFT THE NEXT WEIGHT MOSTLY HORIZONTALLY, SINCE THE BALANCES ARE SHALLOW AND THE DISTANCE MOVED NEED NOT BE VERY GREAT. THESE USUALLY USE A COMPLICATED SYSTEM OF PULLEYS, AND CAN EVEN BE MOUNTED ABOVE AND BELOW ONE ANOTHER. FEW IF ANY ARE KNOWN TO HAVE BEEN BUILT IN 2022, LIKE MOST OF COPPEDGE'S DESIGNS.

- MODULAR LEVERS: IT WAS THOUGHT AFTER AWHILE THAT THERE CAN BE WAYS OF LIFTING MARBLES ONTO A SERIES OF LEVERS WHICH COULD MEAN THAT THE MARBLE COULD GAIN ALTITUDE OVER TIME VERY GRADUALLY.

    - 1ST FULLY PROVABLE: THIS DEVICE THOUGHT OF IN 2016 WAS DESIGNED TO BE BUILT ALONG SIDEWALKS OR AROUND BUILDINGS IN AN URBAN ENVIRONMENT. IT CAN USE VARIOUS RATIOS, TYPICALLY THE LONG END OPERATES A TALL MARBLE, WITH THE COUNTERWEIGHT BEING EVEN HEAVIER. THE MARBLE RISES ALONG A SHALLOW SLOPE, AND THEN USES THE HEIGHT OF IT'S MIDPOINT TO APPLY PRESSURE TO LOWER THE NEXT LEVER SLIGHTLY TO WHERE THE BALL MAY CONTINUE USING THE SPACE PROVIDED BY THE DROP TO CONTINUE MOVING USING THE FOLLOWING LEVER SUCCESSIVELY.

    - CAT-TRAP: A CLEVER DESIGN WHICH MAKES USE OF THE PROPERTIES OF A BALANCE TO CREATE UPWARDS MOTION IN A LIGHTWEIGHT BALL TOWARDS THE FULCRUM RATHER THAN AWAY. NEAR THE FULCRUM, THE BALL IS THEN DEFLECTED SIDEWAYS ONTO ANOTHER SUCH APPARATUS. THOUGHT DIFFICULT, THIS DEVICE SEEMS TO BE PROVEN IN SOME RATIOS, THOUGH IT MAY REQURIE SOME LUCK.

- SEE ALSO WATER-BASED DEVICES.

ESCHER MACHINES

- THE ORIGINAL OR REVERSE ESCHER USES WHAT IS CALLED THE 'MASTER ANGLE' TO MOVE A SINGLE BALL WITH PRESSURE FROM THE 45-DEGREE ANGLED BACKBOARD AND THE POWER OF A VERY SPECIFICALLY-ANGLED WEDGE. THE PARTS MUST BE FIRMLY ATTACHED, AND EVEN A SMALL CHANGE OF ANGLE WILL EASILY PREVENT THE MACHINE FROM RUNNING. HOWEVER, IT CAN USE A WHOLE SERIES OF SPHERES WHICH ALL RUN AUTOMATICALLY INDEPENDENTLY. EXPERIMENTS HAVE BEEN DONE ON THIS SUGGESTING IT WORKS.

- AVANT-GARDE ESCHER MACHINE. BESIDES WORKING WITH VARIOUS TYPES OF WEDGES, THERE IS A VARIATION THAT RUNS USING A LEVER WHICH IS NOT COUNTERWEIGHTED, USING THE FORCE OF THE ESCHER MACHINE TO MOVE A WHEELIE USING NO OTHER ADDED FORCE. THIS IS BASICALLY PROVEN TO WORK, THOUGH IT MIGHT BE DIFFICULT TO BUILD.

- SEE ALSO UNDER UNDERWATER DEVICES.

ESCHER LEVER

- THOUGH NOT A TYPICAL ESCHER MACHINE, THE ESCHER LEVER IS SO-NAMED BECAUSE IT CLEVERLY PERMITS A BALL TO RISE UPWARD ALONG AN UPWARD-DIRECTED LEVER BY USING PRESSURE FROM A COUNTERWEIGHT AND ALSO A SPECIALLY-CURVED SIDEWAYS SUPPORT DURING THE UPWARD MOTION. THOUGH EXTREMELY HARD TO ARRANGE, WITH A SUFFICIENTLY-LIGHTWEIGHT PLATFORM TWISTED IN JUST THE RIGHT WAY, AND HELPED ON ONE SIDE BY AN ADDITIONAL 'WALL' WHICH IS SPECIALLY-CURVED TO BEGIN STEEP AND END SHALLOW BUT SLIGHTLY HIGHER, AND WITH A VERY SHALLOW APPROACH, THE BALL WILL IN SOME CASES TRAVEL AUTOMATICALLY UPWARDS AND EVEN BE ABLE TO RETURN ALONG THE DOWNWARDS SLOPE NEXT TO ITS ORIGINAL PATH, HENCE THE NAME ESCHER LEVER.

BALLOON-BASED

- BALLOON 1: A DEVICE USING A TOP BALLOON APPLYING 0.65X BUOYANCY HAS BEEN FOUND TO POSSIBLY LIFT A SMALL DANGLING WEIGHT WITH THE ABILITY FOR THE DANGLING WEIGHT TO DRAG DOWN THE BALLOON ONCE A DROP IS REACHED.

- BALLOON 2: A SIMILAR DEVICE USING 0.65X DANGLING MASS COMPARED TO BUOYANCY HAS BEEN FOUND TO WORK WITH THE BALLOON SUPPORTED FROM ABOVE WITH THE BALLOON MOVING FIRST DOWNWARDS, THEN LIFTING THE DANGLING WEIGHT WHEN IT REACHES THE END OF THE UPPER SUPPORT. THE UPWARD SUPPORT MAY THEN BE CONTINUED IN A ZIG-ZAG PATTERN HORIZONTALLY. A VERSION OF THIS THAT MAY WORK BEST IS USING VERY SMALL BALLOONS WITH EVEN SMALLER WEIGHTS TO MAXIMIZE THE EFFECT WITHOUT REQUIRING LARGE AMOUNTS OF HELIUM.

FLYING MACHINES

- IT WAS FOUND THE DEVICES MAY FUNCTION JUST AS WELL DESIGNED UPSIDE-DOWN WHEN THERE IS ADDITIONAL BUOYANCY IN THE FORM OF SUPPORT FROM LIGHTER-THAN-AIR BALLOONS, AND WHEN THE WEIGHTS ARE REPLACED WITH THE SAME PROPORTIONS OF BUOYANCY.

ZIPLINE-BASED DEVICES:

- USING SUPPORT VERSUS NON-SUPPORT

- USING DIFFERENCES OF LEVERAGE TO CREATE ALTITUDE ADVANTAGES.

PLUNGE-DEVICES

- THE NIBW6 WAS DESIGNED TO LIFT A HEAVIER WEIGHT ALONG A SPIRAL USING A LONG COUNTER-LEVER ON THE OPPOSITE END, WITH THE SPIRAL BEING FIXED, AND THE BALL BEING LIFTED BY A PLATE OR OTHER SHAPE ATTACHED TO THE LEVER ON THE SHORT END, PASSING THROUGH THE CENTER OF THE SPIRAL, ALLOWING THE BALL TO BE LIFTED AND THEN APPLY PRESSURE IN A SECTION WHERE THE SPIRAL DOES NOT PROVIDE SUPPORT.

- ANOTHER DEVICE I CALL THE PLUNGE LEVER USES A VERY HEAVY COUNTERWEIGHT AT SHORT DISTANCE, APPLYING PRESSURE TO A MEDIUM-LENGTH LEVER ON THE OPPOISTE END, TO WHICH IS ATTACHED A COLUMN-LIKE ELEMENT WHICH PASES INTO A TALL VERTICAL SPIRAL TRACK, WHICH IS FIXED. THE DEVICE IS DESIGNED SO THAT A BALL CAN BE LIFTED BY A PLATE-LIKE ATTACHMENT ON THE TOP OF THE COLUMN, PUSHED UPWARD BY THE HEAVILY COUNTERWEIGHTED COLUMN, AND THEN BE DEFLECTED AT THE TOP TO A POSITION OF VERY SLIGHTLY GREATER LEVERAGE (DUE TO RADIUS) ON THE OUTSIDE OF THE COLUMN. BECAUSE THE LONGER END STILL HAS LEVERAGE AT THE TOP, THIS WOULD CAUSE THE COUNTERWEIGHT TO BE LIFTED IN CERTAIN RATIOS, AND FOR THE BALL TO DESCEND, STILL AT THE

TOP OF THE PLATE. AT THE BASE IT COULD BE DEFLECTED INWARDS, ALLOWING IT TO BE LIFTED AGAIN.

MAGNETS

- A RUSSIAN DEVICE SEEMS TO CLAIM MAGNETS ARE USED TO LIFT WEIGHTS UP SLIGHT INCLINES AND DROP BY THEIR OWN PRESSURE, IF THIS IS WORKABLE IT IS A GOOD IDEA.

- IT IS THOUGHT MAGNETS CAN BE USED TO LIFT WEIGHT UPWARD ALONG SLANTED SURFACES, THE PROBLEM WITH THIS BEING THAT THE MAGNETS DO NOT ALWAYS RELEASE IF THE PRESSURE IS ENOUGH TO LIFT THE WEIGHT. PERHAPS THE POTENTIAL HERE IS WITH A HORIZONTALLY SUPPORTED CART WHICH HAS VARIOUS DEGREES OF LEVERAGE AND MAGNETISM.

- IT HAS BEEN FOUND MAGNETS DON'T HAVE STRONG ADVANTAGES ON COUNTERWEIGHTS, AND CAN BE RELATIVELY EXPENSIVE.

THE CORE SYSTEM

UNIFIED PROOF THEORY

THE MOTIVATION FOR COHERENCE

"My sense is when someone really points out an epistemic weakness by and large this is because a particular claim was held to the standard of a particular field or domain. Really, without the assumptions native to ANY field, there is no point in finding a claim epistemically useless. The strength of a claim, then, is not when a claim is un-falsifiable, OR either when a claim is not questioned, but rather when there is no point in finding a claim epistemically useless, which occurs in inverse proportion to the number of fields inside (and potentially, in philosophy, outside) science, which consider the claim worth questioning. [Coherence, then, implies making use of categories. Categories, for example, like physics or linguistics]." —Problems with Un-Falsifiability (… Coherence Criticism)

…

SYSTEMIC ARGUMENT FOR NATHAN COPPEDGE'S CATEGORICAL DEDUCTION

(1) Every term has an opposite that can be imagined.

(2) Opposites have associations.

(3) Opposites can exist in a bounded Cartesian Coordinate System. (This is meant simply for organization of true opposites like material versus abstract, not arbitrary ones that share a common category).

(4) Opposites must be opposed along the diagonal, because it is the furthest possible distance.

(5) Only opposites are contradictory.

(6) Therefore, terms that are non-opposite can be compared.

(7) Each opposite pair measures coherence, since it has an infinite span.

(8) True opposites must remain in opposite positions during comparisons.

(9) Contingent relations alternate while retaining opposites in opposite positions.

(10) Exponential efficiency (fewer deductions than categories).

(11) Optionally, it may be important to use a language in which states and qualities are roughly scientifically equivalent, since this makes sentence-forming easier.

—<u>Can philosophy be axiomatized?</u>

...

CATEGORICAL DEDUCTION: A LUCKY DISPOSABLE METHOD:

(Resulting from the proof...)

Relatively coherent deduction: "AB:CD and AD:CB" in terms of A.

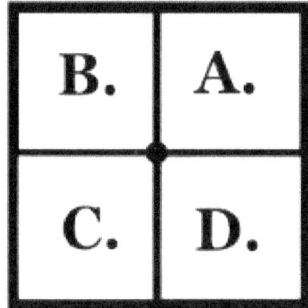

...

THE 11 SYSTEMS: DEFENDING CATEGORICAL DEDUCTION AGAINST ALTERNATES:

1. COHERENT MODELS

- In the incomplete coherent model, if it fails either there is not yet a complete model, or the most complete model may fail to be coherent.
- In the incomplete coherent model, if it succeeds, then various independent parts of science are adequate to describe multiple independent fields without unification.
- In the complete coherent model if it fails either coherence itself is transcended through equal yet good systems, or it is found technology cannot fulfill every part of coherence.
- In the complete coherent model if it succeeds, then absolutely everything is described by one theory, or every possible technology can be predicted.

2. NAIVE REALISM

- Naive sets, full, failed: If sets are full but there was no effort to show extreme efficiency, then there is no risk but nothing is proven.
- Naive sets, full, successful: If sets are full but there was no effort to show efficiency and that is seen as a good thing, that shows lack of foresight.
- Nave sets, empty, failed: If sets are empty but the effort was to show extreme efficiency, then a failure to show efficiency suggests less efficiency.
- Naive sets, empty, successful: If sets are empty and extreme efficiency is proven, then success with efficiency suggests new extreme forms of logic.

3. PARADOXES

- Paradoxes, problematic, un-successful, suggests problems of various kinds.
- Paradoxes, problematic, successful, suggests problems can be solved.
- Paradoxes, un-problematic, un-successful, suggests that the case was not problematic or was not paradoxical.
- Paradoxes, un-problematic, successful, suggests problems or paradoxes have been solved

4. IRRATIONALITY

- Failing to oppose reason, discontinuously. Doesn't seem to have an opinion.
- Failing to oppose reason, continuously. Dominated by an un-emotional force.
- Succeeding to oppose reason, discontinuously. Capable of making statements.
- Succeeding to oppose reason, continuously. Having diabolical power.

5. INCOHERENCE

- General knowledge that is un-impressive. Too general.
- General knowledge that is impressive. Good for education.
- Special knowledge that is un-impressive. Not too technical.
- Special knowledge that is impressive. Actually useful.

6. NEUTRALS

- Not actually neutral, un-important, then not generally important
- Not actually neutral, important, then important for non-neutrals.
- Actually neutral, un-important, then true neutral.
- Actually neutral, important, then important for neutrals.

7. INFORMALS

- Un-appealing, non-instructive. Inconvenient or messy.
- Appealing, non-instructive. Childish and crude.
- Un-appealing, instructive. Boring, dense
- Appealing, instructive. Elaborate, organized.

8. 7TH CATEGORY EXCEPTION

- Relative, absolute: can be relativized again.
- Relative, relative, relativized again.
- Relatively relative, absolute, already absolute.
- Relatively relative, relative: exists as measurement.

9. ACCEPTING RELATIVISM

- By argument, absolute: then relativism must try to be useful.
- By argument, relative: then there must be some way to measure.
- By opinion, absolute: then you should accept suffering as if it's you're own opinion or you are just lucky.
- By opinion, relative: then you don't believe pain is real, or are arguing over opinions not truth.

10. NONSENSICALS

- Uncertain uncertainty. Extreme doubt, may suggest not grasping the picture or needing mental help.
- Certain uncertainty. Confusion.
- Certain certainty. Doubting doubtfulness, yet more doubting.
- Uncertain certainty. Assuming questionable foundations.

11. IMPOSSIBLES

- Possible possibles: could still be impossible.
- Possible impossibles, could still be possible.
- Impossible impossibles, could be more possible than impossible.
- Impossible matchik, could make impossible impossibles possible.

...

PAROXYSM; GENERAL SOLUTION TO PARADOXES AND PROBLEMS:

- Find the best definition of the problem. Select the polar opposite of EVERY WORD in the best definition of the original problem (except the word 'paradox' which is not included), and combine them in the same order as the corresponding original words.

Paradoxes must be solvable, because we know they are either problematic or non-problematic. If they are non-problematic, they are solvable or unsolvable. If they are unsolvable, they are problems, not paradoxes. Thus, paradoxes are either solvable and problematic or solvable and non-problematic, therefore there must be an opposite of a paradox for every paradox which solves it. Applied to coherent problems, it is a natural fit that opposites express anything equivalent to 'problematic' and 'non-problematic' (though ambiguously as to which one, in coherence, the poles remain in opposite positions, and thus, there is still logical relation). Thus, half of a coherent diagram will express a problem, and the other half must express a solution, or it may be assumed non-paradoxical or incoherent (judging by how non-opposites will not express all extremes).

—General Solution to All Problems and Paradoxes (...)

...

FORMULA FOR THE SOULS OF BOOKS

Title of book = '[quality of X] [opp qualifier]'

Soul of the book = 'If you [X] qualifier [subject of X and qualifier] [opp X clarified]'

Socrates observed that the soul is ironic. In a snap-judgment I observed that Socrates meant a formula for the soul. From this I could conclude the formula involved contradiction. I concluded the second element must be the contradiction, and the soul must have a name, and the soul must not contradict itself, so the second element must contradict the name, and if the first element of the name is the quality, and quality is nature, and nature means psyche, then if the name has two elements the second part of the soul contradicts the second part of the name. The third part of the soul must not contradict further, so it is the result or conjunction of the first two parts, and if there is a fourth element it must prove the soul by attempting to contradict the first element, which requires clarity.

—Formula for Souls (...)

APPLYING CATEGORICAL DEDUCTION TO FORMULA FOR SOULS:

From a document called 'Nathan Coppedge Computers':

- GIVEN QUESTION (A) is C: BD THEN BCAD and / or DCAB
- GIVEN QUESTION (B) is D: CA THEN CDBA and / or ADBC
- GIVEN QUESTION (C) is A: DB THEN DACB and / or BACD
- GIVEN QUESTION (D) is B: AC THEN ABDC and / or CBDA

Given: A. Abstract, B. Systems, C. Physical, D. Foundations (where A,B,C,D can be word labels, and we assume C is polar opposite of A, and D is polar opposite of B):

Quick knowledge according to Categorical Deduction: {AB:CD and AD:CB} results in: Abstract Systems Physical Foundations, and Abstract Foundations Physical Systems

The question (A) is C: B-D, for example, Q (Abstract) = physics: systems-foundations

The question (B) is D: C-A, for example, Q (Systems) = foundations: physical-abstract

The question (C) is A: D-B, for example, Q (Physical) = abstract: foundations-systems

The question (D) is B: A-C, for example, Q (Foundations) = systems: abstract-physical

Soul Formula: Noun for (Adj 1 to Opp Adj 1 summarized) + Opp of (Noun 1 to Opp Noun 1 summarized) + (Result of Noun and Opposite together) + Opposite of same Noun = Meaning (Translation).

The meaning of A question is BCAD or DCAB. Meaning of (A) is systems founded in abstract foundations, or foundations of physical abstract systems.

The meaning of B question is CDBA or ADBC. Meaning of (B) is physical foundations of systemic abstractions, or abstract foundations of systematic physics.

The meaning of C question is DACB or BACD. Meaning of (C) is foundational abstractions for physical systems, or system of abstract physical foundations.

The meaning of D question is ABDC or CBDA. Meaning of (D) is abstract systems founded in physics, or physical systems founded in abstractions.

—<u>Nathan Coppedge Computers</u> (…)

UNIVERSE DNA: AN INTERMEDIATE STEP TO THE THEORY OF EVERYTHING

BCAD or DCAB AND (CDBA or ADBC) AND DACB or BACD AND (ABDC or CBDA)

OR

BCAD or DCAB AND (ABDC or CBDA) AND DACB or BACD AND (CDBA or ADBC)

As shown at: Universe DNA (GOOGLE DOCS) this is simply the four solutions from Nathan Coppedge Computers with Categorical Deduction applied to the order of the categories, assuming a specified set order (which is potentially arbitrary, but does not involve an arbitrary interpretation). This is an attempt at a complete universal formula without much mathematical insight, as I realized.

...

RESULTS FORMULA (TOE)

- Everything is examples.
- Examples might have useful results, otherwise they are different. And they may have differences even if they have useful results.
- A useful result might be translated as an efficiency.
- Efficiencies are different than differences.
- Since something without efficiency could be expressed as zero efficiency, and something with positive efficiency could express a technology, the formula may be expressed as Efficiency + Difference.
- However, this may not be precise enough. If we search for nothing, there is no efficiency (no effect) and the difference is what remains.
- If the efficiency is greater than the difference, if there is at least one result, then if the difference is >0, then the efficiency will be >1, because so far we are assuming the efficiency is a positive number, so when we add the efficiency and the difference we get results, and we are assuming at least one result.
- Since efficiency is only <= 1 in a closed energy system, we can conclude an open system involves efficiency > 1, and all other cases have an efficiency from < 1 to > 0, since so far we are assuming efficiency is positive.
- Thus the formula becomes qualified. Expressed in limits we get:

- Set 0 > Efficiency* + Difference where efficiency sums to < 1 if topic is acted on (that is, less than unity), and efficiency sums to > 1 if topic is acting (that is, greater than unity).—<u>The Shortest Proof of the TOE</u> (...)

ANTITHEORY OF EVERYTHING (REVERSING)

Because mathematical and coherent, Antitheory of Everything (Results) can be taken as Reverse of sign and reverse of contents of TOE (Results Formula).

TOE: Results >= Efficiency + Difference

Anti-Thing (Results) <= Difference - Efficiency

—<u>The Antitheory</u> (...)

...

DIFFERENCE FORMULA

It has been shown by convention that a difference of 1 means unity. Intuitively this represents a system which has zero energy losses, and no definition could be more conservative in regards to perfect systems.

Sensibly, from a difference of 1, we can subdivide and determine that a difference of 0.5 is normal for positive energy which is not over-unity, and -0.5 is normal for negative energies which are not hyper-negative. We can then surmise the existence of a negative energy of 1. Since the positive energies correspond with physical efficiency, the negative energy must refer to things such as antimatter and abstraction.

We can then form the beginnings of a function spectrum:

-1 fully abstract, -0.5 abstract, 0 neutral, +0.5 material, +1 unity

Since there is no principle of ending a number line with ordinal numbers, we must assume that the function spectrum is theoretically-potentially infinite unless the entire universe is finite.

—<u>Difference Theories</u> (...)

EFFICIENCY FORMULA

With two of three variables defined, the Efficiency formula can be determined with simple arithmetic: Results = Efficiency + Difference, so Efficiency = Results - Difference.

See also: Unified Efficiency Theory (...)

DISINTEGRAL FORMULA (MIN # OF PARTS?)

The equation simply takes the antitheory and negates it, to find a more coherent version of the Theory of Everything. Antitheory = Difference - Efficiency. Disintegral = - (Difference - Efficiency)

—Disintegrals (...)

...

COHERENCE EQUATION: SOLUTION TO THE COMPLETE COMPLETENESSS PROBLEM

(This is a means of reaching a string of more coherent unified formulas. However, it ultimately results in a repetition of the Disintegral and minus zero, whatever minus zero means is unclear).

This is the same method that was applied to the TOE to reach the disintegral, one of the only things more coherent than the TOE. Apply the Antitheory and negate. Apply the Antitheory and negate (a formula for making a unified theory more coherent)...

TOE: Coherence degree 1: Efficiency + Difference

Disintegral: Coherence degree 2: - (Difference - Efficiency)

Coherence Degree 3: - [-(Difference - Efficiency) + (Difference - Efficiency)]

Coherence Degree 4: -{-[-(Difference - Efficiency) + (Difference - Efficiency)] + (Difference - Efficiency)} ... And etc it keeps going.

DEGREE <= -1: Incoherence.

DEGREE 0: Art.

DEGREE 1: Results >= Efficiency + Difference

DEGREE 2: Results >= - (Difference - Efficiency) [=Disintegral]

DEGREE 3: Zero or minus zero

DEGREE 4: Disintegral

DEGREE 5: Zero or minus zero

DEGREE 6: Disintegral

DEGREE 7: Zero or minus zero

DEGREE 8: Disintegral

DEGREE 9: Zero or minus zero

DEGREE 10: Disintegral

DEGREE 11: Zero or minus zero

DEGREE 12: Disintegral

—Coherence Equation (...)

...

INFINITE TOWER:

Successful Realization of the Disintegral idea is that the Disintegral formula alternates to zero to infinite degrees in high levels of analysis, with the disintegral occupying even degrees. The exception is degrees zero to one, and negative.

It was later found not all pinnacle theories adopt this pattern, though many of htem use zero and minus zero as one of the repeating increments.

—Coherence Equation (...)

...

COHERENCE AND CHEMICAL REWARDS

The reward for a standard logic can ideally have certain results:

- Descriptive model.
- Continuation of the logic.
- Description of a finite infinite.
- Exponentiation of some external.

These clearly do not provide much of a reward for the user's chemical system.

Let us look at the results in terms of pleasure:

- Description of pleasure: reduction to a description, requires magic.
- Continuation of pleasure: a serotonin cycle or typical process. Typical process might be enhanced but is otherwise un-exciting. Serotonin cycle comes at a cost.
- Description of a finite infinite: an exaggerated description, not authentic pleasure.
- Exponentiation of an external: enhancement of things like vision and noise. This might depend on more advanced technology before being realized except for low-stimulus events.

Now if instead we look at coherence, the rewards for a coherent logic express the following:

- Expression of a complete model.
- Containment of the model.
- Description of the absolute.
- Exponential efficiency.

Translating for pleasure:

- The sense of fulfillment (completeness). Obviously more beneficial than before.
- Containment, which is to say, continuing reward, possibly more advanced than serotonin.
- The absolute, which is to say, ultimate, or a scientific equivalent (perhaps).
- A principle of exponential efficiency, which means an enhanced effect of the above.

Coherence is a better model for chemical rewards than traditional logic.

SUBLIME ENGINEERING BY NATHAN COPPEDGE

GENERAL DEFENSE OF EXPONENTIAL EFFICIENCY

"It coheres automatically is an invention of Nathanian criticism. It has been argued this makes it the best theory. This is because (1) It must only be used if you are being fair, (2) It is disposable, (3) It uses timeless arguments, and (4) It is exponentially efficient."
—NC

PROOF OF EXPONENTIAL EFFICIENCY (THREE SECTIONS BELOW)

It was realized since perpetual motion expressed variable Efficiency, and the TOE expressed variable Results, a third theory would unify them having to do with variable Difference. This had already been found to be the Function Spectrum under the Difference Formula.

- KNOWLEDGE: Results (1,2,3...) = Eff + Difference
- PERPETUAL MOTION: Results = Eff (1,2,3...) + Difference
- FUNCTION SPECTRUM: Results = Eff + Difference (1,2,3...)
- Unified—!

OBJECTIVE KNOWLEDGE

Lists of principles have been taken to mean mathematical axioms.

If we look at mathematics, there is a hidden trend towards generalism.

Yet, the axioms of mathematics do not claim to be a comprehensive theory.

In generalism, if a theory is not comprehensive, it is not successful.

If the axioms are not comprehensive, the list of principles cannot be complete no matter their quantity.

Thus, mathematical principles tend to never succeed as models.

Then, if one wishes to form a comprehensive theory, one should attempt to construct a comprehensive model.

If mathematics is the closest known thing to constituents of such model, then pseudo-mathematical principles might be used comprehensively to create a model that is comprehensive.

Logical relationships could be used in place of mathematical relationships, to express the idea of pseudo-mathematics in a way that is rigorous.

To express a formal context completely, polar opposite relations might be used in order to span any amount of distance.

However, the realization of a comprehensive system depends at the very least on having an advantage on all other (knowledge) contexts.

Since the other contexts mentioned are incomplete models, and we are not yet assuming that the model is complete (if it is thought to be a model), then we must express some efficiency which places the purported comprehensive model above all incoherent (incomplete) models.

Since we decided the comprehensive systems must be pseudo-mathematical in other words, logical, the use of exponents to express efficiency may be the best approach for creating a model which has an advantage against incoherence.

Now, we have decided that what we might call 'coherent knowledge' involves a complete list of categories in which exponential efficiency is created.

Since a complete list of categories is required, but it is unknown how many total opposites exist in the universe, we have two possible approaches: (1) We can create a logic which works for any number of opposites even a small number, or (2) We can relativize our logic so that measurement of relativism is sufficient for absoluteness.

Since we don't want an incomplete model, we must choose both of these options, in a method called relative absoluteness, in which relative relativism produces absoluteness. Relativism is equated with measurement. The measurement can take place by creating a logic which is exponentially efficient.

Since opposites are concepts which tend to represent opposite extremes, it makes sense that if they are actual polar opposites like abstraction versus materialism, they would represent DIAGONALLY opposite extremes, which therefore create only two combinations in four categories. Thus, if for completeness we read all the categories in every combination, there are already only two combinations for four categories, thus the problem of exponential efficiency is solved.

Assuming the simplest combination is the most fundamental, the equation that is produced for a true comprehensive model is: AB:CD and AD:CB also written more generally as: Results (1,2,3...) = Eff + Difference.

(The problem of exponential efficiency is solved coherently and the measurement can take place if the categories are considered to express all they contain = relative relativism. The set remains complete if the categories are treated as equal, resulting in a net-neutral relationship of all categories, which means symmetric enough for the adoption of universal logic over all categories. If the system is not fair in this way, that would suggest non-universal application which would not qualify as a complete model. Obviously incomplete models should be avoided, because they are seen as having less advantageous logic).

Nonetheless, as advantageous as it is compared to scientific models, objective knowledge should be treated as disposable (in principle at least) because if treated as an imperative it could have disastrous consequences when used unfairly. It is not a moral system or an attempt to change human fate, but rather an effort at understanding the universe and 'creating' meaning.

PERPETUAL MOTION

Properties may be quantified (space, matter).

The properties may be SPECIAL as they are quantified matter.

Specialization requires application.

In specialism, if it is not applied, it may fail.

However, energy anticipates variation in heat such as occurs in a Carnot engine, which anticipates variable efficiency, i.e. variable energy conversion rates.

If generalism (the state of the universe) predicts variable efficiency, it contradicts this sort of specialism.

Since efficiency must vary (given that specialism must involve variation by space, matter), specialism must fail.

However, this sort of generalism still needs to generalize over specialism if it is to be a physical model.

Mechanical relationships could be used in place of general relationships, after all, efficiency in this case is mechanical (physical).

Mechanical relationships might be expressed generally as exponential (generalized thus abstract and mathematical, efficient) relationships.

SUBLIME ENGINEERING BY NATHAN COPPEDGE

The goal in general of this sort of specialization might be to express differences in energy.

Since it expresses the potential of all specialized contexts, and we are not assuming that the model is inefficient, what is expressed is a form of ultimate mechanical efficiency.

Since what is being discussed is energy efficiency (physics) and what is needed is a general theory of EFFICIENCY in which energy is not constant, the use of exponents to express differences in energy may be expedient.

Now we have decided that what is called 'efficiency mechanics' involves differences of energy in which exponential efficiency is involved.

Since differences of energy are assumed for every type of efficiency, a combination of efficiencies will be optimal. A combination of efficiencies will involve: (1) Efficiencies which are applied to one another, or (2) efficiencies which do not apply to one another.

Since we are looking for exponential efficiency, it is more likely that the efficiencies apply to one another. Thus, it seems we have found the application involves exponentially-efficient mechanics.

Since the exponentially-efficient mechanics involves differences, it makes sense that optimally the differences will create the exponential efficiency.

Assuming the general formula is the most useful, the simplest explanation is that it takes the form: Results = Eff (1,2,3...) + Difference.

...

FUNCTION SPECTRUM

The combination of quantities and mathematics can be taken to mean functions.

Functions may be general or special.

Since generalized is taken to mean abstract, and specialized is taken to mean material, unifying them will imply creating a continuum between the abstract and material. Since abstract and material are seen to be exclusive, it can be done through their constituent parts: quantities and mathematics, generalism and specialism.

The included theories express the Theory of Anything = Results = Efficiency + Difference.

Since the include contents both express the TOE, and they are exclusive, thus if a continuum is created, the continuum is exclusive.

The continuum must incorporate energy inefficiencies as well as incomplete models. Thus, a unity of differences. Since this unified continuum will not contradict either of these exclusive theories that are part of the theory of anything, it will be non-contradictory.

Since the theory involves unity and over-unity, it will be a theory of energy.

As it adopts the Theory of Anything, it is not pseudo-mathematical: it is actually mathematical, so far as that theory is accurate.

Thus, it is both general and special.

Since it is both general and special, it may express one in terms of the other, creating mutual efficiency, as shown in the theories of objective knowledge and perpetual motion.

The theory must be a successful general and special theory.

As the theory is a general and special theory, it need not assume generalism or specialism.

Since the TOE is general concerning special theories, the continuum also includes incompleteness and inefficiencies.

Now we have decided that what we might call a 'Continuum of Differences' involves mathematics which expresses energy and abstraction.

Since mathematics is required, and material and abstract are seen as opposites, there are two categories of the Function Spectrum: (1) Negative differences, representing abstraction, and (2) Positive differences, representing matter.

Although it expresses the whole TOE, we said it is a continuum of two separate parts, thus positive and negative differences will be assumed to be part of the same infinite axis.

As a continuum is designed to express linear differences, it makes sense that the continuum is measured in terms of constant difference values which occur in perpetual motion and knowledge along particular specific points, reflected in the TOE equation.

Assuming the most obvious combination is the most direct, the resulting differences give initially +1 for perpetual motion, and -1 for knowledge, with the remainder expressed as differences and efficiencies with correlated results according to the TOE. More generally it can be written as: Results = Eff + Difference (1,2,3...)...

—Proof of Exponential Efficiency 2022–02–19

ADVANCED RESEARCH:

TOE / COHERENT TREE

(This is a means of reaching a string of more coherent unified formulas. However, it ultimately results in a repetition of the Disintegral and minus zero, whatever minus zero means is unclear).

This is the same method that was applied to the TOE to reach the disintegral, one of the only things more coherent than the TOE. Apply the Antitheory and negate. Apply the Antitheory and negate (a formula for making a unified theory more coherent)...

TOE: Coherence degree 1: Results (1,2,3...) = Efficiency + Difference

Disintegral: Coherence degree 2: Results (1,2,3...) = - (Difference - Efficiency)

Coherence Degree 3: Results (1,2,3...) = - [-(Difference - Efficiency) + (Difference - Efficiency)]

Coherence Degree 4: Results (1,2,3...) = -{-[-(Difference - Efficiency) + (Difference - Efficiency)] + (Difference - Efficiency)} ... And etc it keeps going.

DEGREE <= -1: Results (1,2,3...) = Incoherence.

DEGREE 0: Results (1,2,3...) = Art.

DEGREE 1: Results (1,2,3...) = Efficiency + Difference [= General Theory of Everything]

DEGREE 2: Results (1,2,3...) = - (Difference - Efficiency) [=Disintegral]

DEGREE 3: Results (1,2,3...) = Zero or minus zero

DEGREE 4: Results (1,2,3...) = Disintegral

DEGREE 5: Results (1,2,3...) = Zero or minus zero

DEGREE 6: Results (1,2,3...) = Disintegral

DEGREE 7: Results (1,2,3...) = Zero or minus zero

DEGREE 8: Results (1,2,3...) = Disintegral

DEGREE 9: Results (1,2,3...) = Zero or minus zero

DEGREE 10: Results (1,2,3...) = Disintegral

DEGREE 11: Results (1,2,3...) = Zero or minus zero

DEGREE 12: Results (1,2,3...) = Disintegral

...

TOE / PERPETUAL MOTION TREE

TOE: Degree 1: Results = Efficiency (1,2,3...) + Difference

Disintegral: Degree 2: Results = - (Difference - Efficiency (1,2,3...))

Degree 3: Results = - [-(Difference - Efficiency (1,2,3...)) + (Difference - Efficiency (1,2,3...))]

Degree 4: Results = -{-[-(Difference - Efficiency (1,2,3...)) + (Difference - Efficiency (1,2,3...))] + (Difference - Efficiency (1,2,3...))} ... And etc it keeps going.

DEGREE <= -1: Results = Incoherence (1,2,3...).

DEGREE 0: Results = Art (1,2,3...).

DEGREE 1: Results = Efficiency (1,2,3...) + Difference [= General Perpetual Motion]

DEGREE 2: Results = - (Difference - Efficiency (1,2,3...)) [= Perpetual Disintegral]

DEGREE 3: Results = Zero or minus zero (1,2,3...)

DEGREE 4: Results = Perpetual Disintegral

DEGREE 5: Results = Zero or minus zero (1,2,3...)

DEGREE 6: Results = Perpetual Disintegral

DEGREE 7: Results = Zero or minus zero (1,2,3...)

DEGREE 8: Results = Perpetual Disintegral

DEGREE 9: Results = Zero or minus zero (1,2,3...)

DEGREE 6: Results = Perpetual Disintegral

DEGREE 7: Results = Zero or minus zero (1,2,3...)

DEGREE 12: Results = Perpetual Disintegral

...

TOE / FUNCTION SPECTRUM TREE

TOE: Degree 1: Results = Efficiency + Difference (1,2,3...)

Disintegral: Degree 2: Results = - (Difference (1,2,3...) - Efficiency)

Degree 3: Results = - [-(Difference (1,2,3...) - Efficiency) + (Difference (1,2,3...) - Efficiency)]

Degree 4: Results = -{-[-(Difference (1,2,3...) - Efficiency) + (Difference (1,2,3...) - Efficiency)] + (Difference (1,2,3...) - Efficiency)} ... And etc it keeps going.

DEGREE <= -1: Results = Incoherence (1,2,3...).

DEGREE 0: Results = Art (1,2,3...).

DEGREE 1: Results = Efficiency + Difference (1,2,3...) [= General Function Spectrum]

DEGREE 2: Results = - (Difference (1,2,3...) - Efficiency) [= Function Spectrum Disintegral]

DEGREE 3: Results = Zero or minus zero (1,2,3...)

DEGREE 4: Results = Function Spectrum Disintegral

DEGREE 5: Results = Zero or minus zero (1,2,3...)

DEGREE 6: Results = Function Spectrum Disintegral

DEGREE 7: Results = Zero or minus zero (1,2,3...)

DEGREE 8: Results = Function Spectrum Disintegral

DEGREE 9: Results = Zero or minus zero (1,2,3...)

DEGREE 6: Results = Function Spectrum Disintegral

DEGREE 7: Results = Zero or minus zero (1,2,3...)

DEGREE 12: Results = Function Spectrum Disintegral

...

ANTI-THEORY RESULTS (ANTI-THEORY KNOWLEDGE) TREE

(This is a means of reaching a string of less coherent, less unified formulas. However, it ultimately results in a repetition of the Anti-Disintegral and zero, whatever zero means is unclear).

This is the same method that was applied to the TOE to reach the disintegral, one of the only things more coherent than the TOE. Apply the TOE and negate. Apply the TOE and negate (a formula for making an incoherent theory more incoherent)...

Anti-Theory: Anti-Coherence degree 1: Results (1,2,3...) = Difference - Efficiency = Antitheory

Anti-Coherence degree 2: Results (1,2,3...) = - (Efficiency + Difference) = Negative TOE (not sure if this means Superdisintegral)

Anti-Coherence Degree 3: Results (1,2,3...) = - (-Efficiency + -Difference + Efficiency + Difference) = 0 or -0

Anti-Coherence Degree 4: Results (1,2,3...) = -(0 + Efficiency + Difference) = Negative TOE

Anti-Coherence Degree 5: - (-Efficiency + -Difference + Efficiency + Difference) = 0 or -0

... And etc it keeps going.

DEGREE <= -1: Antitheory (1,2,3...) = Incoherence. (?)

DEGREE 0: Antitheory (1,2,3...) = Art. (?)

DEGREE 1: Antitheory (1,2,3...) = Difference - Efficiency [= Knowledge of the Antitheory]

DEGREE 2: Antitheory (1,2,3...) = - (Efficiency + Difference) = Negative TOE

DEGREE 3: Antitheory (1,2,3...) = - (-Efficiency + -Difference + Efficiency + Difference) [= 0 or -0]

DEGREE 4: Antitheory (1,2,3...) = - (Eff + Diff) [= Negative TOE]

DEGREE 5: Antitheory (1,2,3...) = 0 or -0

DEGREE 6: Antitheory (1,2,3...) = Negative TOE

DEGREE 7: Antitheory (1,2,3...) = 0 or -0

DEGREE 8: Antitheory (1,2,3...) = Negative TOE

DEGREE 9: Antitheory (1,2,3...) = 0 or -0

DEGREE 10: Antitheory (1,2,3...) = Negative TOE

DEGREE 11: Antitheory (1,2,3...) = 0 or -0

DEGREE 12: Antitheory (1,2,3...) = Negative TOE

...

ANTI-THEORY EFFICIENCY (ANTI-THEORY PERPETUAL MOTION) TREE

Anti-Theory: Anti-Efficiency degree 1: Results = Difference - Efficiency (1,2,3...) = Perpetual Motion Antitheory

Anti-Perpetual Motion degree 2: Results = - (Efficiency (1,2,3...) + Difference) = Negative TOE (not sure if this means Superdisintegral)

Anti-Perpetual Motion Degree 3: Results = - (-Efficiency (1,2,3...) + -Difference + Efficiency (1,2,3...) + Difference) = 0 or -0

Anti-Perpetual Motion Degree 4: Results (1,2,3...) = -(0 + Efficiency (1,2,3...) + Difference) = Negative TOE

Anti-Perpetual Motion Degree 5: - (-Efficiency (1,2,3...) + -Difference + Efficiency (1,2,3...) + Difference) = 0 or -0

... And etc it keeps going.

DEGREE <= -1: Antitheory = Incoherence (1,2,3...)

DEGREE 0: Antitheory = Art (1,2,3...)

DEGREE 1: Antitheory = Difference - Efficiency (1,2,3...) [= Perpetual Motion Antitheory]

DEGREE 2: Antitheory = - (Efficiency (1,2,3...) + Difference) [= Negative TOE]

DEGREE 3: Antitheory = Difference - Efficiency (1,2,3...) [= Perpetual Motion Antitheory]

DEGREE 4: Antitheory = Negative TOE

DEGREE 5: Antitheory = 0 or -0

DEGREE 6: Antitheory = Negative TOE

DEGREE 7: Antitheory = 0 or -0

DEGREE 8: Antitheory = Negative TOE

DEGREE 9: Antitheory = 0 or -0

DEGREE 10: Antitheory = Negative TOE

DEGREE 11: Antitheory = 0 or -0

DEGREE 12: Antitheory = Negative TOE

...

ANTI-THEORY DIFFERENCE (ANTI-FUNCTION SPECTRUM) TREE

Anti-Theory: Anti-Function degree 1: Results = Difference (1,2,3...) - Efficiency = Anti-Function Spectrum

Anti-Function degree 2: Results = - (Efficiency + Difference (1,2,3...)) = Negative TOE (not sure if this means Superdisintegral)

Anti-Function Degree 3: Results (1,2,3...) = - (-Efficiency + -Difference (1,2,3...) + Efficiency + Difference (1,2,3...)) = 0 or -0

Anti-Function Degree 4: Results (1,2,3...) = -(0 + Efficiency + Difference (1,2,3...)) = Negative TOE

Anti-Function Degree 5: - (-Efficiency + -Difference (1,2,3...) + Efficiency + Difference (1,2,3...)) = 0 or -0

... And etc it keeps going.

DEGREE <= -1: Antitheory = Incoherence. (1,2,3...)

SUBLIME ENGINEERING BY NATHAN COPPEDGE

DEGREE 0: Antitheory = Art. (1,2,3...)

DEGREE 1: Antitheory = Difference (1,2,3...) - Efficiency [= Anti-Function Spectrum]

DEGREE 2: Antitheory = - (Efficiency + Difference (1,2,3...)) = Negative TOE

DEGREE 3: Antitheory = Difference (1,2,3...) - Efficiency [= Anti-Function Spectrum]

DEGREE 4: Antitheory = Negative TOE

DEGREE 5: Antitheory = 0 or -0

DEGREE 6: Antitheory = Negative TOE

DEGREE 7: Antitheory = 0 or -0

DEGREE 8: Antitheory = Negative TOE

DEGREE 9: Antitheory = 0 or -0

DEGREE 10: Antitheory = Negative TOE

DEGREE 11: Antitheory = 0 or -0

DEGREE 12: Antitheory = Negative TOE

...

EFFICIENCY / EFFICIENCY TREE (NOTE: IN THE CASE OF EFFICIENCY, EFFICIENCY IS PLACED AHEAD OF RESULTS-BASED FORMULAS FOR REASONS THAT BECOME OBVIOUS)

(This is a means of reaching a string of more efficient formulas. However, it ultimately results in a repetition).

This is a similar method to applying the TOE to reach the disintegral. In this case, Apply the Anti-Efficiency (addition) and negate. Apply the Anti-Efficiency and negate (a formula for making an efficient theory more efficient)...

Efficiency: Efficiency degree 1: Efficiency (1,2,3...) >= Results – Difference = Efficiency Theory

Efficiency degree 2: Efficiency (1,2,3...) = - (Difference - Results) = Results - Difference = Efficiency Theory (1,2,3...)

Efficiency Degree 3: Efficiency (1,2,3…) = - (Results - Difference + Difference - Results) = 0 or -0

Efficiency Degree 4: Efficiency (1,2,3…) = - (0 + Difference - Results) = Results - Difference = Efficiency Theory …

Efficiency Degree 5: Efficiency (1,2,3…) = - (Results - Difference + Difference - Results) = 0 or -0 … And etc it keeps going.

DEGREE <= -1: Efficiency (1,2,3…) = Incoherence (?)

DEGREE 0: Efficiency (1,2,3…) = Art (?)

DEGREE 1: Efficiency (1,2,3…) = Results – Difference [= Efficiency Theory]

DEGREE 2: Efficiency (1,2,3…) = Results - Difference [= Efficiency Theory]

DEGREE 3: Efficiency (1,2,3…) = 0 or -0

DEGREE 4: Efficiency (1,2,3…) = Efficiency Theory

DEGREE 5: Efficiency (1,2,3…) = 0 or -0

DEGREE 6: Efficiency (1,2,3…) = Efficiency Theory

DEGREE 7: Efficiency (1,2,3…) = 0 or -0

DEGREE 8: Efficiency (1,2,3…) = Efficiency Theory

DEGREE 9: Efficiency (1,2,3…) = 0 or -0

DEGREE 10: Efficiency (1,2,3…) = Efficiency Theory

DEGREE 11: Efficiency (1,2,3…) = 0 or -0

DEGREE 12: Efficiency (1,2,3…) = Efficiency Theory

…

SUBLIME ENGINEERING BY NATHAN COPPEDGE

EFFICIENCY / EFFICIENT KNOWLEDGE TREE

Efficiency: Efficiency degree 1: Efficiency >= Results (1,2,3...) – Difference = Efficient Knowledge

Efficiency degree 2: Efficiency = - (Difference - Results (1,2,3...)) = Results (1,2,3...) - Difference = Efficient Knowledge.

Efficiency Degree 3: Efficiency = - (Results (1,2,3...) - Difference + Difference - Results (1,2,3...)) = 0 or -0

Efficiency Degree 4: Efficiency = - (0 + Difference - Results (1,2,3...)) = Results (1,2,3...) - Difference = Efficient Knowledge ...

Efficiency Degree 5: Efficiency = - (Results (1,2,3...) - Difference + Difference - Results (1,2,3...)) = 0 or -0 ... And etc it keeps going.

DEGREE <= -1: Efficiency = Incoherence (1,2,3...)

DEGREE 0: Efficiency = Art (1,2,3...)

DEGREE 1: Efficiency = Results (1,2,3...) – Difference [= Efficient Knowledge]

DEGREE 2: Efficiency = Results (1,2,3...) – Difference [= Efficient Knowledge]

DEGREE 3: Efficiency = 0 or -0

DEGREE 4: Efficiency = Efficient Knowledge

DEGREE 5: Efficiency = 0 or -0

DEGREE 6: Efficiency = Efficient Knowledge

DEGREE 7: Efficiency = 0 or -0

DEGREE 8: Efficiency = Efficient Knowledge

DEGREE 9: Efficiency = 0 or -0

DEGREE 10: Efficiency = Efficient Knowledge

DEGREE 11: Efficiency = 0 or -0

DEGREE 12: Efficiency = Efficient Knowledge

SUBLIME ENGINEERING BY NATHAN COPPEDGE

EFFICIENCY / EFFICIENT FUNCTION SPECTRUM TREE

Efficiency: Efficiency degree 1: Efficiency >= Results – Difference (1,2,3…) = Efficient Function Spectrum

Efficiency degree 2: Efficiency = - (Difference (1,2,3…) - Results) = Results - Difference (1,2,3…) = Efficient Function Spectrum.

Efficiency Degree 3: Efficiency = - (Results - Difference (1,2,3…) + Difference (1,2,3…) - Results) = 0 or -0

Efficiency Degree 4: Efficiency = - (0 + Difference (1,2,3…) - Results) = Results - Difference (1,2,3…) = Efficient Function Spectrum …

Efficiency Degree 5: Efficiency = - (Results - Difference (1,2,3…) + Difference (1,2,3…) - Results) = 0 or -0 … And etc it keeps going.

DEGREE <= -1: Efficiency = Incoherence (1,2,3…)

DEGREE 0: Efficiency = Art (1,2,3…)

DEGREE 1: Efficiency = Results – Difference (1,2,3…) [= Efficient Function Spectrum]

DEGREE 2: Efficiency = Results - Difference (1,2,3…) [= Efficient Function Spectrum]

DEGREE 3: Efficiency = 0 and -0

DEGREE 4: Efficiency = Efficient Function Spectrum

DEGREE 5: Efficiency = 0 and -0

DEGREE 6: Efficiency = Efficient Function Spectrum

DEGREE 7: Efficiency = 0 and -0

DEGREE 8: Efficiency = Efficient Function Spectrum

DEGREE 9: Efficiency = 0 and -0

DEGREE 10: Efficiency = Efficient Function Spectrum

DEGREE 11: Efficiency = 0 and -0

DEGREE 12: Efficiency = Efficient Function Spectrum

SUBLIME ENGINEERING BY NATHAN COPPEDGE

ANTI-EFFICIENCY / ANTI-EFFICIENCY TREE

(This is a means of reaching a string of anti-efficient formulas. However, it ultimately results in a repetition).

This is a similar method to applying the TOE to reach the disintegral. In this case, Apply Efficiency (addition) and negate. Apply the Efficiency and negate (a formula for making an anti-efficient theory less efficient)...

Anti-Efficiency: Anti-Efficiency degree 1: Anti-Efficiency (1,2,3...) <= Difference - Results = Anti-Efficiency

Anti-Efficiency degree 2: Anti-Efficiency (1,2,3...) <= - (Results - Difference) = Difference - Results = Anti-Efficiency (1,2,3...).

Anti-Efficiency Degree 3: Anti-Efficiency (1,2,3...) = - (Difference - Results + Results - Difference) = 0 or -0

Anti-Efficiency Degree 4: Anti-Efficiency (1,2,3...) = - (0 + Results - Difference) = Difference - Results = Anti-Efficiency... And etc it keeps going.

DEGREE <= -1: Anti-Efficiency (1,2,3...) = Incoherence (?)

DEGREE 0: Anti-Efficiency (1,2,3...) = Art (?)

DEGREE 1: Anti-Efficiency (1,2,3...) = Difference - Results [= Anti-Efficiency]

DEGREE 2: Anti-Efficiency (1,2,3...) = Difference - Results [= Anti-Efficiency]

DEGREE 3: Anti-Efficiency (1,2,3...) = 0 or -0

DEGREE 4: Anti-Efficiency (1,2,3...) = Anti-Efficiency

DEGREE 5: Anti-Efficiency (1,2,3...) = 0 or -0

DEGREE 6: Anti-Efficiency (1,2,3...) = Anti-Efficiency

DEGREE 7: Anti-Efficiency (1,2,3...) = 0 or -0

DEGREE 8: Anti-Efficiency (1,2,3...) = Anti-Efficiency

DEGREE 9: Anti-Efficiency (1,2,3...) = 0 or -0

DEGREE 10: Anti-Efficiency (1,2,3...) = Anti-Efficiency

DEGREE 11: Anti-Efficiency (1,2,3...) = 0 or -0

DEGREE 12: Anti-Efficiency (1,2,3...) = Anti-Efficiency

...

ANTI-EFFICIENCY / ANTI-EFFICIENT KNOWLEDGE TREE

Anti-Efficiency: Anti-Efficiency degree 1: Anti-Efficiency <= Difference - Results (1,2,3...) = Anti-Efficient Knowledge

Anti-Efficiency degree 2: Anti-Efficiency <= - (Results (1,2,3...) - Difference) = Difference - Results (1,2,3...) = Anti-Efficient Knowledge.

Anti-Efficiency Degree 3: Anti-Efficiency = - (Difference - Results (1,2,3...) + Results (1,2,3...) - Difference) = 0 or -0

Anti-Efficiency Degree 4: Anti-Efficiency = - (0 + Results (1,2,3...) - Difference) = Difference - Results (1,2,3...) = Anti-Efficient Knowledge... And etc it keeps going.

DEGREE <= -1: Anti-Efficiency = Incoherence (?)

DEGREE 0: Anti-Efficiency = Art (?)

DEGREE 1: Anti-Efficiency = Difference - Results (1,2,3...) [= Anti-Efficient Knowledge]

DEGREE 2: Anti-Efficiency = Difference - Results (1,2,3...) [= Anti-Efficient Knowledge]

DEGREE 3: Anti-Efficiency = 0 or -0

DEGREE 4: Anti-Efficiency = Anti-Efficient Knowledge

DEGREE 5: Anti-Efficiency = 0 or -0

DEGREE 6: Anti-Efficiency = Anti-Efficient Knowledge

DEGREE 7: Anti-Efficiency = 0 or -0

DEGREE 8: Anti-Efficiency = Anti-Efficient Knowledge

DEGREE 9: Anti-Efficiency = 0 or -0

DEGREE 10: Anti-Efficiency = Anti-Efficient Knowledge

DEGREE 11: Anti-Efficiency = 0 or -0

DEGREE 12: Anti-Efficiency = Anti-Efficient Knowledge

...

ANTI-EFFICIENCY / ANTI-EFFICIENT FUNCTION SPECTRUM TREE

Anti-Efficiency: Anti-Efficiency degree 1: Anti-Efficiency <= Difference (1,2,3...) - Results = Anti-Efficient Function Spectrum

Anti-Efficiency degree 2: Anti-Efficiency <= - (Results - Difference (1,2,3...)) = Difference (1,2,3...) - Results) = Anti-Efficient Function Spectrum.

Anti-Efficiency Degree 3: Anti-Efficiency = - (Difference (1,2,3...) - Results + Results - Difference (1,2,3...)) = 0 or -0

Anti-Efficiency Degree 4: Anti-Efficiency = - (0 + Results - Difference (1,2,3...)) = Difference (1,2,3...) - Results = Anti-Efficient Function Spectrum... And etc it keeps going.

DEGREE <= -1: Anti-Efficiency = Incoherence (?)

DEGREE 0: Anti-Efficiency = Art (?)

DEGREE 1: Anti-Efficiency = Difference (1,2,3...) - Results [= Anti-Efficient Function Spectrum]

DEGREE 2: Anti-Efficiency = Difference (1,2,3...) - Results [= Anti-Efficient Function Spectrum]

DEGREE 3: Anti-Efficiency = 0 or -0

DEGREE 4: Anti-Efficiency = Anti-Efficient Function Spectrum

DEGREE 5: Anti-Efficiency = 0 or -0

DEGREE 6: Anti-Efficiency = Anti-Efficient Function Spectrum

DEGREE 7: Anti-Efficiency = 0 or -0

DEGREE 8: Anti-Efficiency = Anti-Efficient Function Spectrum

DEGREE 9: Anti-Efficiency = 0 or -0

SUBLIME ENGINEERING BY NATHAN COPPEDGE

DEGREE 10: Anti-Efficiency = Anti-Efficient Function Spectrum

DEGREE 11: Anti-Efficiency = 0 or -0

DEGREE 12: Anti-Efficiency = Anti-Efficient Function Spectrum

...

DIFFERENCE / DIFFERENT FUNCTION SPECTRUM TREE

(This is a means of reaching a string of different formulas. However, it ultimately results in a repetition).

This is a similar method to applying the TOE to reach the disintegral. In this case, Apply the Anti-Difference (addition) and negate. Apply the Anti-Difference and negate (a formula for making a difference theory more different)...

Difference: Difference degree 1: Difference (1,2,3...) >= Results − Efficiency = Difference Theory

Difference degree 2: Difference (1,2,3...) = - (Efficiency - Results) = Results - Efficiency = Difference Theory (1,2,3...).

Difference Degree 3: Difference (1,2,3...) = - (Results - Efficiency + Efficiency - Results) = 0 or -0

Difference Degree 4: Difference (1,2,3...) = - (0 + Efficiency - Results) = Results - Efficiency = Difference Theory (1,2,3...) ... And etc it keeps going.

DEGREE <= -1: Difference (1,2,3...) = Incoherence (?)

DEGREE 0: Difference (1,2,3...) = Art (?)

DEGREE 1: Difference (1,2,3...) = Results − Efficiency [= Difference Theory]

DEGREE 2: Difference (1,2,3...) = Results − Efficiency [= Difference Theory]

DEGREE 3: Difference (1,2,3...) = 0 or -0

DEGREE 4: Difference (1,2,3...) = Difference Theory

DEGREE 5: Difference (1,2,3...) = 0 or -0

DEGREE 6: Difference (1,2,3...) = Difference Theory

DEGREE 7: Difference (1,2,3...) = 0 or -0

DEGREE 8: Difference (1,2,3...) = Difference Theory

DEGREE 9: Difference (1,2,3...) = 0 or -0

DEGREE 10: Difference (1,2,3...) = Difference Theory

DEGREE 11: Difference (1,2,3...) = 0 or -0

DEGREE 12: Difference (1,2,3...) = Difference Theory

...

DIFFERENCE / DIFFERENT TOE TREE

Difference: Difference degree 1: Difference >= Results (1,2,3...) – Efficiency = Different Knowledge

Difference degree 2: Difference = - (Efficiency - Results (1,2,3...)) = Results (1,2,3...) - Efficiency = Different Knowledge.

Difference Degree 3: Difference = - (Results (1,2,3...) - Efficiency + Efficiency - Results (1,2,3...)) = 0 or -0

Difference Degree 4: Difference = - (0 + Efficiency - Results (1,2,3...)) = Results (1,2,3...) - Efficiency = Different Knowledge ... And etc it keeps going.

DEGREE <= -1: Difference = Incoherence (1,2,3...)

DEGREE 0: Difference = Art (1,2,3...)

DEGREE 1: Difference = Results (1,2,3...) – Efficiency [= Different Knowledge]

DEGREE 2: Difference = - (Results (1,2,3...) – Efficiency + Efficiency - Results (1,2,3...)) [= 0 or - 0]

DEGREE 3: 0 or -0

DEGREE 4: Difference = Different Knowledge

DEGREE 5: 0 or -0

DEGREE 6: Difference = Different Knowledge

SUBLIME ENGINEERING BY NATHAN COPPEDGE

DEGREE 7: 0 or -0

DEGREE 8: Difference = Different Knowledge

DEGREE 9: 0 or -0

DEGREE 10: Difference = Different Knowledge

DEGREE 11: 0 or -0

DEGREE 12: Difference = Different Knowledge

...

DIFFERENCE / DIFFERENT PERPETUAL MOTION TREE

Difference: Difference degree 1: Difference >= Results – Efficiency (1,2,3...) = Different Perpetual Motion

Difference degree 2: Difference = - (Efficiency (1,2,3...) - Results) = Results - Efficiency (1,2,3...) = Different Perpetual Motion.

Difference Degree 3: Difference = - (Results - Efficiency (1,2,3...) + Efficiency (1,2,3...) - Results) = 0 or -0

Difference Degree 4: Difference = - (0 + Efficiency (1,2,3...) - Results) = Results - Efficiency (1,2,3...) = Different Perpetual Motion... And etc it keeps going.

DEGREE <= -1: Difference = Incoherence (1,2,3...)

DEGREE 0: Difference = Art (1,2,3...)

DEGREE 1: Difference = Results – Efficiency (1,2,3...) [= Different Perpetual Motion]

DEGREE 2: Difference = Results – Efficiency (1,2,3...) [= Different Perpetual Motion]

DEGREE 3: Difference = 0 or -0

DEGREE 4: Difference = Different Perpetual Motion

DEGREE 5: Difference = 0 or -0

DEGREE 6: Difference = Different Perpetual Motion

DEGREE 7: Difference = 0 or -0

DEGREE 8: Difference = Different Perpetual Motion

DEGREE 9: Difference = 0 or -0

DEGREE 10: Difference = Different Perpetual Motion

DEGREE 11: Difference = 0 or -0

DEGREE 12: Difference = Different Perpetual Motion

...

ANTI-DIFFERENCE / ANTI-DIFFERENCE TREE

(This is a means of reaching a string of anti-difference formulas. However, it ultimately results in a repetition).

This is a similar method to applying the TOE to reach the disintegral. In this case, Apply Difference (addition) and negate. Apply the Difference and negate (a formula for making an anti-difference theory less different)...

Anti-Difference: Anti-Difference degree 1: Anti-Difference (1,2,3...) <= Efficiency - Results = Anti-Difference

Anti-Difference degree 2: Anti-Difference (1,2,3...) <= - (Results – Efficiency) = Efficiency - Results = Anti-Difference (1,2,3...).

Anti-Difference Degree 3: Anti-Difference (1,2,3...) = - (Efficiency - Results + Results – Efficiency) = 0 or -0

Anti-Difference Degree 4: Anti-Difference (1,2,3...) = - (0 + Results – Efficiency) = Efficiency - Results = Anti-Difference ... And etc it keeps going.

DEGREE <= -1: Anti-Difference (1,2,3...) = Incoherence (?)

DEGREE 0: Anti-Difference (1,2,3...) = Art (?)

DEGREE 1: Anti-Difference (1,2,3...) = Efficiency - Results [= Anti-Difference]

DEGREE 2: Anti-Difference (1,2,3...) = Efficiency - Results [= Anti-Difference]

DEGREE 3: Anti-Difference (1,2,3...) = 0 or -0

DEGREE 4: Anti-Difference (1,2,3...) = Anti-Difference

DEGREE 5: Anti-Difference (1,2,3...) = 0 or -0

DEGREE 6: Anti-Difference (1,2,3...) = Anti-Difference

DEGREE 7: Anti-Difference (1,2,3...) = 0 or -0

DEGREE 8: Anti-Difference (1,2,3...) = Anti-Difference

DEGREE 9: Anti-Difference (1,2,3...) = 0 or -0

DEGREE 10: Anti-Difference (1,2,3...) = Anti-Difference

DEGREE 11: Anti-Difference (1,2,3...) = 0 or -0

DEGREE 12: Anti-Difference (1,2,3...) = Anti-Difference

...

ANTI-DIFFERENCE / ANTI-DIFFERENT TOE TREE

Anti-Difference: Anti-Difference degree 1: Anti-Difference <= Efficiency - Results (1,2,3...) = Anti-Different TOE

Anti-Difference degree 2: Anti-Difference <= - (Results (1,2,3...) – Efficiency) = Efficiency - Results (1,2,3...) = Anti-Different TOE.

Anti-Difference Degree 3: Anti-Difference = - (Efficiency - Results (1,2,3...) + Results (1,2,3...) – Efficiency) = 0 or -0

Anti-Difference Degree 4: Anti-Difference (1,2,3...) = - (0 + Results – Efficiency) = Efficiency - Results = Anti-Different TOE ... And etc it keeps going.

DEGREE 0: Anti-Difference = Art (1,2,3...)

DEGREE 1: Anti-Difference = Efficiency - Results (1,2,3...) [= Anti-Different TOE]

DEGREE 2: Anti-Difference = Efficiency - Results (1,2,3...) [= Anti-Different TOE]

DEGREE 3: Anti-Difference = 0 or -0

DEGREE 4: Anti-Difference = Anti-Different TOE

DEGREE 5: Anti-Difference = 0 or -0

DEGREE 6: Anti-Difference = Anti-Different TOE

DEGREE 7: Anti-Difference = 0 or -0

DEGREE 8: Anti-Difference = Anti-Different TOE

DEGREE 9: Anti-Difference = 0 or -0

DEGREE 10: Anti-Difference = Anti-Different TOE

DEGREE 11: Anti-Difference = 0 or -0

DEGREE 12: Anti-Difference = Anti-Different TOE

...

ANTI-DIFFERENCE / ANTI-DIFFERENT PERPETUAL MOTION TREE

Anti-Difference: Anti-Difference degree 1: Anti-Difference <= Efficiency (1,2,3...) - Results = Anti-Different Efficiencies

Anti-Difference degree 2: Anti-Difference <= - (Results – Efficiency (1,2,3...)) = Efficiency (1,2,3...) - Results = Anti-Different Efficiencies.

Anti-Difference Degree 3: Anti-Difference = - (Efficiency (1,2,3...) - Results + Results – Efficiency (1,2,3...)) = 0 or -0

Anti-Difference Degree 4: Anti-Difference = - (0 + Results – Efficiency (1,2,3...)) = Efficiency (1,2,3...) - Results = Anti-Different Efficiencies ... And etc it keeps going.

DEGREE <= -1: Anti-Difference = Incoherence (1,2,3...)

DEGREE 0: Anti-Difference = Art (1,2,3...)

DEGREE 1: Anti-Difference = Efficiency (1,2,3...) - Results [= Anti-Different Efficiencies]

DEGREE 2: Anti-Difference = Efficiency (1,2,3...) - Results [= Anti-Different Efficiencies]

DEGREE 3: Anti-Difference = 0 or -0

DEGREE 4: Anti-Difference = Anti-Different Efficiencies

DEGREE 5: Anti-Difference = 0 or -0

DEGREE 6: Anti-Difference = Anti-Different Efficiencies

DEGREE 7: Anti-Difference = 0 or -0

DEGREE 8: Anti-Difference = Anti-Different Efficiencies

DEGREE 9: Anti-Difference = 0 or -0

DEGREE 10: Anti-Difference = Anti-Different Efficiencies

DEGREE 11: Anti-Difference = 0 or -0

DEGREE 12: Anti-Difference = Anti-Different Efficiencies

...

FORCES / FORCES TREE

(This is a means of reaching a string of unified forces. However, it ultimately results in a repetition).

This is the same method that was applied to the TOE to reach the disintegral, one of the only things more coherent than the TOE. Apply the Antiforce and negate. Apply the Antiforce and negate (a formula for generating a string of unified forces)...

Forces: # Forces (1,2,3...) = # Dimensions - # Antiforces = Force Theory

Forces degree 2: Forces (1,2,3...) = - (# Dimensions - # Forces (1,2,3...)) = # Negative Antiforces (correct).

Forces Degree 3: Forces (1,2,3...) = - (# Forces (1,2,3...) - # Dimensions + # Dimensions - # Forces (1,2,3...)) = 0 or -0

Forces Degree 4: - (0 + # Dimensions - # Forces (1,2,3...)) = # Negative Antiforces... And etc it keeps going.

DEGREE <= -1: Forces (1,2,3...) = Incoherence (?)

DEGREE 0: Forces (1,2,3...) = Art (?)

DEGREE 1: Forces (1,2,3...) = # Dimensions - # Antiforces [= Force Theory]

DEGREE 2: Forces (1,2,3...) = - (# Dimensions - # Forces (1,2,3...)) [= # Negative Antiforces]

DEGREE 3: Forces (1,2,3...) = 0 or -0

DEGREE 4: Forces (1,2,3...) = Negative Antiforces

DEGREE 5: Forces (1,2,3...) = 0 or -0

DEGREE 6: Forces (1,2,3...) = Negative Antiforces

DEGREE 7: Forces (1,2,3...) = 0 or -0

DEGREE 8: Forces (1,2,3...) = Negative Antiforces

DEGREE 9: Forces (1,2,3...) = 0 or -0

DEGREE 10: Forces (1,2,3...) = Negative Antiforces

DEGREE 11: Forces (1,2,3...) = 0 or -0

DEGREE 12: Forces (1,2,3...) = Negative Antiforces

...

FORCES / FORCES DIMENSIONS TREE

Forces: # Forces = # Dimensions (1,2,3...) - # Antiforces = Force Dimensions

Forces degree 2: Forces = - (# Dimensions (1,2,3...) - # Forces) = # Negative Antiforces (correct).

Forces Degree 3: Forces = - (# Forces - # Dimensions (1,2,3...) + # Dimensions - # Forces) = 0 or -0

Forces Degree 4: - (0 + # Dimensions (1,2,3...) - # Forces) = # Negative Antiforces... And etc it keeps going.

DEGREE <= -1: Forces = Incoherence (1,2,3...)

DEGREE 0: Forces = Art (1,2,3...)

DEGREE 1: # Forces = # Dimensions (1,2,3...) - # Antiforces [= Force Dimensions]

DEGREE 2: - (# Dimensions (1,2,3...) - # Forces) [= # Negative Antiforces]

DEGREE 3: Forces = 0 or -0

DEGREE 4: Forces = Negative Antiforces

DEGREE 5: Forces = 0 or -0

DEGREE 6: Forces = Negative Antiforces

DEGREE 7: Forces = 0 or -0

DEGREE 8: Forces = Negative Antiforces

DEGREE 9: Forces = 0 or -0

DEGREE 10: Forces = Negative Antiforces

DEGREE 11: Forces = 0 or -0

DEGREE 12: Forces = Negative Antiforces

...

FORCES / FORCES OF ANTIFORCE TREE

Forces: # Forces = # Dimensions - # Antiforces (1,2,3...) = Force of Antiforce Theory

Forces degree 2: Forces = - (# Dimensions - # Forces) = # Negative Antiforces (1,2,3...) (correct).

Forces Degree 3: Forces = - (# Forces - # Dimensions + # Dimensions - # Forces) = 0 or -0

Forces Degree 4: - (0 + # Dimensions - # Forces) = # Negative Antiforces (1,2,3...) ... And etc it keeps going.

DEGREE <= -1: Forces = Incoherence (1,2,3...)

DEGREE 0: Forces = Art (1,2,3...)

DEGREE 1: # Forces = # Dimensions - # Antiforces (1,2,3...) [= Force of Antiforce Theory]

DEGREE 2: Forces = - (# Dimensions - # Forces) = # Negative Antiforces (1,2,3...)

DEGREE 3: Forces = 0 or -0

DEGREE 4: Forces = Negative Antiforces

DEGREE 5: Forces = 0 or -0

DEGREE 6: Forces = Negative Antiforces

DEGREE 7: Forces = 0 or -0

DEGREE 8: Forces = Negative Antiforces

DEGREE 9: Forces = 0 or -0

DEGREE 10: Forces = Negative Antiforces

DEGREE 11: Forces = 0 or -0

DEGREE 12: Forces = Negative Antiforces

...

ANTIFORCES / ANTIFORCE THEOREM TREE

Antiforces: # Antiforces (1,2,3...) = # Dimensions - # Forces = Antiforce Theory

Antiforces degree 2: Antiforces = - (# Dimensions - #Antiforces (1,2,3...)) = # Negative Forces (correct).

Antiforces Degree 3: Antiforces = - (#Antiforces - # Dimensions + # Dimensions - # Antiforces) = 0 or -0

Antiforces Degree 4: - (0 + # Dimensions - # Antiforces) = # Negative Forces (1,2,3...) ... And etc it keeps going.

DEGREE <= -1: Antiforces (1,2,3...) = Incoherence

DEGREE 0: Antiforces (1,2,3...) = Art

DEGREE 1: Antiforces (1,2,3...) = # Dimensions - # Forces [= Antiforce Theory]

DEGREE 2: Antiforces = - (# Dimensions - #Antiforces (1,2,3...)) = # Negative Forces

DEGREE 3: Antiforces = 0 or -0

DEGREE 4: Antiforces = Negative Forces

DEGREE 5: Antiforces = 0 or -0

DEGREE 6: Antiforces = Negative Forces

DEGREE 7: Antiforces = 0 or -0

DEGREE 8: Antiforces = Negative Forces

DEGREE 9: Antiforces = 0 or -0

DEGREE 10: Antiforces = Negative Forces

DEGREE 11: Antiforces = 0 or -0

DEGREE 12: Antiforces = Negative Forces

...

ANTIFORCES / DIMENSIONS OF ANTIFORCES TREE

Antiforces: # Antiforces = # Dimensions (1,2,3...) - # Forces = Dimensions of Antiforces

Antiforces degree 2: Antiforces = - (# Dimensions (1,2,3...) - #Antiforces) = # Negative Forces (correct).

Antiforces Degree 3: Antiforces = - (#Antiforces - # Dimensions (1,2,3...) + # Dimensions (1,2,3...) - # Antiforces) = 0 or -0

Antiforces Degree 4: - (0 + # Dimensions (1,2,3...) - # Antiforces) = # Negative Forces ... And etc it keeps going.

DEGREE <= -1: Antiforces = Incoherence (1,2,3...)

DEGREE 0: Antiforces = Art (1,2,3...)

DEGREE 1: Antiforces = # Dimensions (1,2,3...) - # Forces [= Dimensions of Antiforces]

DEGREE 2: Antiforces = - (# Dimensions (1,2,3...) - #Antiforces) = # Dimensions of Negative Force

DEGREE 3: Antiforces = 0 or -0

DEGREE 4: Antiforces = Negative Forces

DEGREE 5: Antiforces = 0 or -0

DEGREE 6: Antiforces = Negative Forces

DEGREE 7: Antiforces = 0 or -0

DEGREE 8: Antiforces = Negative Forces

DEGREE 9: Antiforces = 0 or -0

DEGREE 10: Antiforces = Negative Forces

DEGREE 11: Antiforces = 0 or -0

DEGREE 12: Antiforces = Negative Forces

...

ANTIFORCES / ANTIFORCES OF FORCE TREE

Antiforces: # Antiforces = # Dimensions - # Forces (1,2,3...) = Antiforces of Force

Antiforces degree 2: Antiforces = - (# Dimensions - #Antiforces) = # Negative Forces (1,2,3...) (correct).

Antiforces Degree 3: Antiforces = - (#Antiforces - # Dimensions + # Dimensions - # Antiforces) = 0 or -0

Antiforces Degree 4: - (0 + # Dimensions - # Antiforces) = # Negative Forces (1,2,3...) ... And etc it keeps going.

DEGREE <= -1: Antiforces = Incoherence (1,2,3...)

DEGREE 0: Antiforces = Art (1,2,3...)

DEGREE 1: Antiforces = # Dimensions - # Forces (1,2,3...) [= Antiforces of Force]

DEGREE 2: Antiforces = - (# Dimensions - #Antiforces) = Negative Forces (1,2,3...)

DEGREE 3: Antiforces = 0 or -0

DEGREE 4: Antiforces = Negative Forces

DEGREE 5: Antiforces = 0 or -0

DEGREE 6: Antiforces = Negative Forces

DEGREE 7: Antiforces = 0 or -0

DEGREE 8: Antiforces = Negative Forces

DEGREE 9: Antiforces = 0 or -0

DEGREE 10: Antiforces = Negative Forces

DEGREE 11: Antiforces = 0 or -0

DEGREE 12: Antiforces = Negative Forces

...

DIMENSIONS / DIMENSIONS TREE

Dimensions: # Dimensions (1,2,3...) = # Forces + # Antiforces = # Dimensions

Dimensions degree 2: Dimensions (1,2,3...) = - (#Antiforces - # Forces) = Negative Anti-Dimensions (1,2,3...) (correct).

Dimensions Degree 3: Antiforces = - (# Forces - # Antiforces + # Antiforces - # Forces) = 0 or -0

Dimensions Degree 4: - (0 + - #Antiforces - # Forces) = Negative Anti-Dimensions (1,2,3...) ... And etc it keeps going.

DEGREE <= -1: Dimensions (1,2,3...) = Incoherence

DEGREE 0: Dimensions (1,2,3...) = Art

DEGREE 1: Dimensions (1,2,3...) = # Forces + # Antiforces [= # Dimensions (1,2,3...)]

DEGREE 2: Dimensions = - (#Antiforces - # Forces) = Negative Anti-Dimensions (1,2,3...)

DEGREE 3: Dimensions = 0 or -0

DEGREE 4: Dimensions = Negative Anti-Dimensions

DEGREE 5: Dimensions = 0 or -0

DEGREE 6: Dimensions = Negative Anti-Dimensions

DEGREE 7: Dimensions = 0 or -0

DEGREE 8: Dimensions = Negative Anti-Dimensions

DEGREE 9: Dimensions = 0 or -0

DEGREE 10: Dimensions = Negative Anti-Dimensions

DEGREE 11: Dimensions = 0 or -0

DEGREE 12: Dimensions = Negative Anti-Dimensions

...

DIMENSIONS / # OF FORCE DIMENSIONS TREE

Dimensions: # Dimensions = # Forces (1,2,3...) + # Antiforces = # Force Dimensions

Dimensions degree 2: Dimensions = - (#Antiforces - # Forces (1,2,3...)) = Negative Anti-Dimensions of Force (1,2,3...) (correct).

Dimensions Degree 3: Antiforces = - (# Forces (1,2,3...) - # Antiforces + # Antiforces - # Forces (1,2,3...)) = 0 or -0

Dimensions Degree 4: - (0 + - #Antiforces - # Forces (1,2,3...)) = Negative Anti-Dimensions (1,2,3...) ... And etc it keeps going.

DEGREE <= -1: Dimensions = Incoherence (1,2,3...)

DEGREE 0: Dimensions = Art (1,2,3...)

DEGREE 1: Dimensions = # Forces (1,2,3...) + # Antiforces [= # Force Dimensions]

DEGREE 2: Dimensions = - (#Antiforces - # Forces (1,2,3...)) [= Negative Anti-Dimensions of Force]

DEGREE 3: Dimensions = 0 or -0

DEGREE 4: Dimensions = Negative Anti-Dimensions of Force

DEGREE 5: Dimensions = 0 or -0

DEGREE 6: Dimensions = Negative Anti-Dimensions of Force

DEGREE 7: Dimensions = 0 or -0

DEGREE 8: Dimensions = Negative Anti-Dimensions of Force

DEGREE 9: Dimensions = 0 or -0

DEGREE 10: Dimensions = Negative Anti-Dimensions of Force

DEGREE 11: Dimensions = 0 or -0

DEGREE 12: Dimensions = Negative Anti-Dimensions of Force

—Coherence Equation (...)

...

DIMENSIONS / # OF ANTIFORCE DIMENSIONS TREE

Dimensions: # Dimensions = # Forces + # Antiforces (1,2,3...) = # Antiforce Dimensions

Dimensions degree 2: Dimensions = - (#Antiforces (1,2,3...) - # Forces) = Negative Anti-Dimensions of Antiforce (1,2,3...) (correct).

Dimensions Degree 3: Antiforces (1,2,3...) = - (# Forces - # Antiforces (1,2,3...) + # Antiforces (1,2,3...) - # Forces) = 0 or -0

Dimensions Degree 4: - (0 + - #Antiforces (1,2,3...) - # Forces) = Negative Anti-Dimensions ... And etc it keeps going.

DEGREE <= -1: Dimensions = Incoherence (1,2,3...)

DEGREE 0: Dimensions = Art (1,2,3...)

DEGREE 1: Dimensions = # Forces + # Antiforces (1,2,3...) [= # Antiforce Dimensions]

DEGREE 2: Dimensions = - (#Antiforces (1,2,3...) - # Forces) [= Negative Anti-Dimensions of Antiforce]

DEGREE 3: Dimensions = 0 or -0

DEGREE 4: Dimensions = Negative Anti-Dimensions of Antiforce

DEGREE 5: Dimensions = 0 or -0

DEGREE 6: Dimensions = Negative Anti-Dimensions of Antiforce

DEGREE 7: Dimensions = 0 or -0

DEGREE 8: Dimensions = Negative Anti-Dimensions of Antiforce

DEGREE 9: Dimensions = 0 or -0

DEGREE 10: Dimensions = Negative Anti-Dimensions of Antiforce

DEGREE 11: Dimensions = 0 or -0

DEGREE 12: Dimensions = Negative Anti-Dimensions of Antiforce

...

ANTI-DIMENSIONS / ANTI-DIMENSIONS TREE

Anti-Dimensions: # Anti-Dimensions (1,2,3...) = # Antiforces - # Forces = # Anti-Dimensions

Anti-Dimensions degree 2: Anti-Dimensions (1,2,3...) = - (# Forces + # Anti-Forces) = Negative Dimensions (correct).

Anti-Dimensions Degree 3: Anti-Dimensions (1,2,3...) = - (- # Anti-Forces - # Forces + # Forces + # Anti-Forces) = 0 or - 0

Anti-Dimensions Degree 4: - (0 + # Forces + # Antiforces) = Negative Dimensions ... And etc it keeps going.

DEGREE <= -1: Anti-Dimensions (1,2,3...) = Incoherence

DEGREE 0: Anti-Dimensions (1,2,3...) = Art

DEGREE 1: Anti-Dimensions (1,2,3...) = # Antiforces - # Forces [= # Anti-Dimensions]

DEGREE 2: Anti-Dimensions (1,2,3...) = - (# Forces + # Anti-Forces) = Negative Dimensions

DEGREE 3: Anti-Dimensions (1,2,3...) = 0 or -0

DEGREE 4: Anti-Dimensions (1,2,3...) = Negative Dimensions

DEGREE 5: Anti-Dimensions (1,2,3...) = 0 or -0

DEGREE 6: Anti-Dimensions (1,2,3...) = Negative Dimensions

DEGREE 7: Anti-Dimensions (1,2,3...) = 0 or -0

DEGREE 8: Anti-Dimensions (1,2,3...) = Negative Dimensions

DEGREE 9: Anti-Dimensions (1,2,3...) = 0 or -0

DEGREE 10: Anti-Dimensions (1,2,3...) = Negative Dimensions

DEGREE 11: Anti-Dimensions (1,2,3...) = 0 or -0

DEGREE 12: Anti-Dimensions (1,2,3...) = Negative Dimensions

...

ANTI-DIMENSIONS / ANTI-DIMENSIONS OF ANTIFORCE TREE

Anti-Dimensions: # Anti-Dimensions = # Antiforces (1,2,3...) - # Forces = # Anti-Dimensions of Antiforce

Anti-Dimensions degree 2: Anti-Dimensions = - (# Forces + # Antiforces (1,2,3...)) = Negative Dimensions of Antiforce.

Anti-Dimensions Degree 3: Anti-Dimensions = - (- # Antiforces (1,2,3...) - # Forces + # Forces + # Antiforces (1,2,3...)) = 0 or - 0

Anti-Dimensions Degree 4: - (0 + # Forces + # Antiforces (1,2,3...)) = Negative Dimensions of Antiforce ... And etc it keeps going.

DEGREE <= -1: Anti-Dimensions = Incoherence

DEGREE 0: Anti-Dimensions = Art

DEGREE 1: Anti-Dimensions = # Antiforces (1,2,3...) - # Forces [= # Anti-Dimensions of Antiforce]

DEGREE 2: Anti-Dimensions = - (# Forces + # Anti-Forces (1,2,3...)) = Negative Dimensions of Antiforce

DEGREE 3: Anti-Dimensions = 0 or -0

DEGREE 4: Anti-Dimensions = Negative Dimensions of Antiforce

DEGREE 5: Anti-Dimensions = 0 or -0

DEGREE 6: Anti-Dimensions = Negative Dimensions of Antiforce

DEGREE 7: Anti-Dimensions = 0 or -0

DEGREE 8: Anti-Dimensions = Negative Dimensions of Antiforce

DEGREE 9: Anti-Dimensions = 0 or -0

DEGREE 10: Anti-Dimensions = Negative Dimensions of Antiforce

DEGREE 11: Anti-Dimensions = 0 or -0

DEGREE 12: Anti-Dimensions = Negative Dimensions of Antiforce

...

ANTI-DIMENSIONS / ANTI-DIMENSIONS OF FORCE TREE

Anti-Dimensions: # Anti-Dimensions of Force = # Antiforces - # Forces (1,2,3...) = # Anti-Dimensions of Force

Anti-Dimensions degree 2: Anti-Dimensions = - (# Forces (1,2,3...) + # Antiforces) = Negative Dimensions of Force (correct).

Anti-Dimensions Degree 3: Anti-Dimensions = - (- # Antiforces - # Forces (1,2,3...) + # Forces (1,2,3...) + # Antiforces) = 0 or - 0

Anti-Dimensions Degree 4: - (0 + # Forces (1,2,3...) + # Antiforces) = Negative Dimensions of Force ... And etc it keeps going.

DEGREE <= -1: Anti-Dimensions = Incoherence

DEGREE 0: Anti-Dimensions = Art

DEGREE 1: Anti-Dimensions = # Antiforces - # Forces (1,2,3...) [= # Anti-Dimensions of Force]

DEGREE 2: Anti-Dimensions = - (# Forces (1,2,3...) + # Antiforces) = Negative Dimensions of Force

DEGREE 3: Anti-Dimensions = 0 or -0

DEGREE 4: Anti-Dimensions = Negative Dimensions of Force

DEGREE 5: Anti-Dimensions = 0 or -0

DEGREE 6: Anti-Dimensions = Negative Dimensions of Force

DEGREE 7: Anti-Dimensions = 0 or -0

DEGREE 8: Anti-Dimensions = Negative Dimensions of Force

DEGREE 9: Anti-Dimensions = 0 or -0

DEGREE 10: Anti-Dimensions = Negative Dimensions of Force

DEGREE 11: Anti-Dimensions = 0 or -0

DEGREE 12: Anti-Dimensions = Negative Dimensions of Force

—NC Tree Theory (...)

BACKGROUNDS:

THIS IS THE NECESSARY PERFECTIONS, MATCHIK HEURISTICS WAS RELOCATED TO THE EARLIER SECTION ON SPELLS.

AMBIANCE

You may be able to improve other's mood at will, or create a general appearance of cheerfulness, or create spirits which have good influence.

…

…

ARCOLOGIES

"They must not be slaves if they live in the Utopia."

"If votes really counted, life would be different. We would live a Fantasy."

"Arcologies are a means for securing everlasting happiness."

AMERICA

Marie Antoinette (formerly Ann Bolyn) supports the formation of America, borrowing money leading to the collapse of the French monarchy, simultaneously inspiring the beginning of common fashion, in effect modernism. She loves the idea of inventing perpetual motion machines and time-travels to escape hell.

—<u>What is the history, origin, and future of man?</u>

THE ARTESIAN WELL

There was a time of magic in which a god made to fashion a well which would join two sublime regions by one Plutonian channel; It was rumored nymphs played there, during seasonal festivals;

Labeled upon the rock were these words:

> Names and nomes art Riddled bare

There was afterwards a time of war and conflict during which one ruling emperor fastened his realm;

The stone had been replaced and refurbished:

> It read: There art names that are Known in hell

Finally, there was an age of forgetfulness;

But someone had been careful to replace the stone; It had a new inscription:

> Some things blesst Are made of sand

Little account is made of the well after that period, for its location has been forgotten

—From my book '1-Page Classics'

BEST OF MARIE ANTOINETTE

- **What she wrote on complexia:** Arcane authentic historical method: It is what precedes psychology. It is at least the seed of psychology. It is the general-general psychology. How to be open-minded.
- **What she wrote on psychology:** The classic example is gilt. It is a paradigm in psüche. One desires something one does not have. It is complex for society. Is it trivial to research? Well, not so much! You see, these 'conditionalities' shape our society, and they are psycho.
- **Her desire to invent perpetual motion:** She noticed none of them were moving. Do any of them move? She asked. None, he said, except by wind. But, could it be a perpetuum mobilis? She asked. The Gardener replied: No, no. You see, it is a typical thing. In my understanding perpetual motion was one of the few things Marie Antoinette would cry about—and she did so copiously for what may have been 6 hours.
- **Soul of Humanity:** You see, Marie Antoinette crafted the human soul out of gilt… It was her proviso… the soul of humanity must be the soul of automata… the soul of automata must be the soul of humanity. You see, the emotions must be weakened, or they die… You say the soul may have died. It was dying of emotion. How are we to believe you. It lives in the future. How? Is this a dream? It is a dream becoming ever more real! They will kill us soon!
- **The Cloissone Egg:** In my understanding perpetual motion was one of the few things Marie Antoinette would cry about—and she did so copiously for what may have been 6 hours. I remember busting a Cloissone egg over that one, and being disappointed that all it contained was money. One can imagine the archetypal Cloissone egg is inscribed with the words: "Sad on rare occasions".
- **Complex argument:** Let's say she did become a miserable wretch, but the miserable wretch made full use of complexity like no one else ever would. Now she is forced to be a wretch, because the wretch used her work, although they did away with her because she was a miserable wretch. Now who is more of a devil, the noble who was called a devil, or the wretch who paid for the devil who was called the devil before, but who now is a miserable wretch? How would she pay for her soul if she was not a wretch who could pay the devil to pay himself?
- **Mysteries:** Perhaps she had a self-playing harpsichord which moved its own keys. This would explain her reputation as a devil. Perhaps she was much stupider than people think, and could not really play harpsichord, or not that well.
- **Occult:** She may have had a secret desire to create perpetual motion as an ornament for her garden which is considered occult. Where did I miss my magics?

BEST OF NATHAN COPPEDGE

- *There's only a handful of forms / On the faces of the strangers / The people laugh and they shake their heads / They tell me there's no danger…*
- SUPPOSEDLY GREATER: I visited an odd shop, I was traveling with my father and brother. A Chinese man with an animated face walked up to me. As he walked, he grasped a machine in his hands. And as he walked, the most wonderful thing began to happen. I was awe struck. Is it a perpetual motion machine? I said. No. He said mysteriously. I was not disappointed. What is it? I said. My Dad said: A self-beating drum.
- MEETS THE CLEVEREST MAN: We were on the beach at Westville, where my step-mother's mother used to live. While we were navigating the rocks, a vision appeared on the horizon.
  A vision of a man walking, and laughing. "I bet you don't know what my secret is?" He said. What it appeared to be was a fly-wheel, with a listening box underneath. The flywheel appeared to be in motion! I analyzed it. Then I listed to it. "It's a fabled perpetual motion machine!" I said. "That's right!" he said. "But you still don't know my secret!" he said. So I busted it on the sidewalk! Inside, I found: A plastic disk designed to look like a flywheel in motion! And, a miniature conch shell, designed to create the sound of the flywheel! "Now, you pay me for that!" he said. I instinctively leapt away. Then, begrudgingly, I pulled out $41. "It's worth at least that much, but that's all I can afford to pay" I said. "I tried to pay you, that's all I can do. Your secret is all I need to invent perpetual motion myself, plus some other things."
- DREAMS OF LIGHTNING: I was in a dark wood during a storm when I thought: "If lightning strikes me I'm dead." This had the tone of a eureka because the idea that I was the inventor of perpetual motion was not so different from the lightning that would supposedly kill me. Was I made of lightning, or was the lightning made of me? Either way, how would it kill me? Was someone saying I would die of genius? But genius is evolution, and evolution has already evolved by definition. And I wept that I hadn't invented perpetual motion, and I envied the genius struck by lightning… Shazzam! I said to myself, and laughed a sad laugh. Now I am lightning. Eureka! I got it wrong! I said to myself, and I cried in my sleep.
- A single path contradicts / With birds flying / or shot by arrows…
- PERPETUAL MOTION SHOPKEEPER'S SPEECH: Oh, that? That's expensive for a reason, my friend. It's a free market, you can imitate one if you want. It's a perpetual motion machine believe it or not. I say you can build it, but it's not as easy as it looks. If you don't design it meeting certain criteria, it will always fail. It's not the devil, it's a real invention. They didn't happen until recently for some reason. They scared the scientists. They still do. Basically it's just genuine mechanics. It's no more complicated than a shoe.
- *While they were Floundering, I was Pondering: No More wandering through the Dark Tunnels of Grim Determination! For No! It is time to grow in a thousand-folded folds, for which we need an Infinite Fuel!*

BEST OF ZHENG GUO

- **SCRIPT:** The first time I achieved the unity of the Five Chinese Elements I also achieved foolishness, and went to hell.

- **INVENTION MONEY:** When he walked through the town carrying coins in buckets from his shoulders, people said 'fortunate man, fortunate man' and took many of the coins away. The town became known for 'fortunate men' who were so rich they could eat much rice.

- **INVENTION WAR:** When soldiers are slow they carry long sticks. The proper response is fancy hands. Fancy hands was like magic. The other soldiers would have been smart to play dead. So the great Kwang won in all combat. Finally he became the new emperor's servant, and served him in battle. The opposing armies learned one lesson: death. The great Kwang was victorious and immortal. Know something perverse. Revel in it. Become a hero. Dance the dance of life. Discover death.

- **INVENTION DEATH:** Sure enough, with enough effort the sorcerer died. Guo felt he gained his power. The gods were never sure what the power was. But they came to call it Death. Death was originally a magic power, it could give immortal life. But when Guo poisoned the sorcerer, the secret was incomplete. Guo consulted dragons on the advice of the emperor, but he would not tell the sad secrets the dragons told. And so was sealed the bargain to contain the misery of heaven.

- **INVENTION NUMBERS:** Guo: You again. Lao: Another might be greater. Guo: The second thing might always be greater. Lao: The second thing might always be wiser. Guo: The second thing might not be God.

- **INVENTION POISON:** The instruction was an opposite of an opposite, a clever invention that was impossible to understand. Because, although it meant the emperor shouldn't take it, it did not mean that he SHOULD take it. And he might think, if the emperor SHOULDN'T take it, then SOMEONE SHOULD. But in fact the meaning was that no one should take it. This puzzle became known to the emperor as king's food, and became known to the villagers as emperor's soup. But only Guo knew the truth. But he did not know his friend was dead.

- **INVENTION GOD:** I remember making elaborate appeals to the Sorceror that the truest way was to die. I would come by numerous times showing him tricks. Sometimes miraculous cures, sometimes invisibility or time-travel. Anything that might be stuck in a god's hat. I couldn't always pull it off, but often I did. I tried to convince him. And it seemed that it worked. There was another argument, that he was weary of the world. But I felt endless mirth about this. It seeemed like the funniest, most dreadful thing. For a god to disappear—poof! I think that was what it was about.

- **INVENTION IMPOSSIBLES:** These were knots the gods would make Impossible that it's impossible. Not right that it's not right. Everything that it is nothing. Subtle that it's subtle. Strange that it's strange.

BOOKS

Soon Yee invented the bound book because she wanted to put her sewing skills to work while traveling on a Chinese junk selling silk around 200 AD.

THE CAKE

People say there is always something wrong with the cake. When you don't need the cake, is when the idealist might win. It might even be that what we call a cake isn't a cake, and a real cake isn't called a cake.

(THE) COHERENT BRAIN

Re-assessment of human ideas:

1. Physical luck. 14. Regular luck.

2. Greed for works. 5. Greed for ideas. 7. Regular greed.

3. Sufficient ideas. 6. Idea. 10. Obviated idea.

4. And 9. Mental sensations.

5. (Skip).

6. (Skip).

7. (Skip).

8. Mental-physical works.

9. (Skip).

10. (Skip).

11. And 12 Physical art.

12. (Skip).

13. Madness.

14. (Skip).

CLEVEREST MAN

We were on the beach at Westville, where my step-mother's mother used to live. While we were navigating the rocks, a vision appeared on the horizon.
A vision of a man walking, and laughing. "I bet you don't know what my secret is?" He said. What it appeared to be was a fly-wheel, with a listening box underneath. The flywheel appeared to be in motion! I analyzed it. Then I listed to it. "It's a fabled perpetual motion machine!" I said. "That's right!" he said. "But you still don't know my secret!" he said. So I busted it on the sidewalk! Inside, I found: A plastic disk designed to look like a flywheel in motion! And, a miniature conch shell, designed to create the sound of the flywheel! "Now, you pay me for that!" he said. I instinctively leapt away. Then, begrudgingly, I pulled out $41. "It's worth at least that much, but that's all I can afford to pay" I said. "I tried to pay you, that's all I can do. Your secret is all I need to invent perpetual motion myself, plus some other things."

CLOCK OF HYPER-CUBISM

"My recent theory is that the traipse of thought is the clock of Hyper-Cubism. Imagine thought lives on the surface of dark metallic lines that lie somewhere between... mathematics and... [the] enchanted... Do we not see that additional time elapses there? That herbal cultures grow there? That thought survives there?"

—<u>**What is thought itself?**</u> (...)

THE CLOISSONE EGG

In my understanding perpetual motion was one of the few things Marie Antoinette would cry about—and she did so copiously for what may have been 6 hours. I remember busting a Cloissone egg over that one, and being disappointed that all it contained was money.

Check yourself before you wreck yourself may have been invented that time.

One can imagine the archetypal Cloissone egg is inscribed with the words:

"Sad on rare occasions".

Marie Antoinette viewed Cloissone eggs as a great mystery, like a door to another world, a message from God, or the spoil of the world turning against her.

CLONES

You too? I asked. There was another man who looked like me. Then I remembered, they were my 'doubles'. I had hired them to give speeches around the country. Your who? Someone asked me. He looked very important. My… cloenes! I said. They're my special men, who do special work for me. You wouldn't understand. Clowns, he said? No, cloenes, it's a French word. Clownes, I'm sure that's not French he said. Or maybe it is. What did you call them? Cloh-Nies, I said. The man laughed. That came off rather badly. At any rate, they're my stunt-doubles.

…

THE DABBLER

Dumb atom.

Deep archaic time.

Dabbling.

A brilliant infographic.

A mysterious tunnel full of Glyphs.

Arachnid computer.

A blanched desert, with blue, winking stars.

A Trap-Door spider.

Terminality and Time-Travel.

Metaphysics.

Dimensions.

The World.

Immortality.

Perpetual Motion.

…

DETECTING HIDDEN ROOMS

Elements + Sets - Number of Known Rooms

THE DREAM OF THE MACHINE

"Do you dream of a machine? The ultimate wish. The ultimate use of symbols. The ultimate machine. The ultimate kind of life. Are you dreaming? Are you dreaming of a machine?" —<u>The Ideal Sense</u> (...)

ELECTRONIC ISLANDS

Metaphysics of Blessings and Curses:

1. Some islands are fake and free of authentic curses.

2. Some islands are really electronic, and have fake curses.

3. Some islands look like electronics, and have electronic curses.

4. Some islands are not islands and have hidden curses.

5. Some islands are shameful, and have bad blessings.

6. Some islands are rich and have typical problems.

7. Some islands are poor, with authentic difficulty.

8. Some islands are a trap, and mean only philosophy.

9. Some islands are real, and truly meaningless.

10. Some places are empty, and few have wandered there.

11. Some places are futile, and entirely empty.

12. Some islands are good, and a dragon sleeps there.

ESSENCIA COMPLEXIA

Essencia Complexia —The crowning achievement of the Dimensional Modality

EULER, A HUMAN DEVIL

Euler was a kind of devil whose name rhymes with 'lucifer'. His entire life was rational and he died as a drug addict in an expensive government program. This is the only past-life I remember where I could summon flames.

FLEUR-DE-LIS

Even the queen of France did not know what the fleur-de-lis meant. Possibly nymphomania.

FOOLISH GENIUSES

Marie Antoinette

Perhaps she had a self-playing harpsichord which moved its own keys. This would explain her reputation as a devil. Perhaps she was much stupider than people think, and could not really play harpsichord, or not that well.

Nathan Coppedge

Nathan Coppedge realized there were so many geniuses in the world that he would do better to play the fool, if only he could take advantage of the most diabolical exceptions. How better to do this than to play the role of a stealthy philosopher in a small college town? Philosophers were just the things that seemed like fools to the scientific geniuses. Yet philosophers were supposed to possess wisdom, so this showed a key blind spot.

Zheng Guo

The first time I achieved the unity of the Five Chinese Elements I also achieved foolishness, and went to hell. As he escaped from the very same cave where he was buried, he took the form of a bird who said: "If I don't know how you know, I might think differently."

…

FORTUNATE TRUTH

I think it's sometimes used to refer to Fortunate Man, Dragon's Treasure, Emperor's Soup, Cloissone Egg, Treasure Bird, Perpetual Motion…

FUTURE-ORIENTED AND PAST-ORIENTED (LOGIC)

Especially diabolical thinkers can think of an 'enfolded idea' such as an 'enfolded history'.

Abstraction — it was discovered by Orchyrae, who blamed Jesus for the burning of the Library of Alexandria. A greater occultation.

Today we would not consider anyone before AD 0 as making a big discovery, as they were 'future oriented' the same way we are 'past oriented' concerning the Library of Alexandria. This is why Orchyrae really lived around AD 0, not in the BCs.

THE GOOD ROBOT

Do not be destroyed.

GREATNESS

I have thought that no person in history except perhaps God has achieved more than 2.5 / 3 things, those things being Extreme Wealth, Unquestioned Popularity, and Utter Originality. Humans rarely if ever achieve more than two out of the three properties of Wealth, Popularity, and Originality. Achieving 2.5 properties has happened in a small number of cases which involved cheating: for example, Jesus became famous and obscure, Henry Ford became a poor rich boy, and Nathan Coppedge became a stupid genius.

Wealth OF 0 to 1, Input, CAN BE CHANGED

Popularity 0 to 1, Input, CAN BE CHANGED

Originality 0 to 1, Input, CAN BE CHANGED

Total, [=SUM(Wealth, Popularity, Originality)], DEPENDENT

"Genius Game Level", [=((TOTAL/2.5)*100)]

"Greatness IQ" —->, [=(((TOTAL / 2) * 120) +40)]

IMAGINATION

Imagination is beyond the level of 'possible giant possible real man'.

As Great Equalizer: Someone can be more evolved by having imagination, even if they imagine something that's primitive like flipper bones.

IMMORTALITY, INSPIRATION FOR

He carried [the mandrake root] through the house, saying it would not kill.

IMMORTALITY, TYPES OF

Soul immortality:

- Soul of immortality.
- Perpetual motion magic (weak: See: <u>The Mysteries of Perpetual Motion</u>)
- Essence of immortality.
- Out-thinking soul.
- Party favors.

Perpetual motion immortality:

- Immortal soul and perpetual motion.
- Perpetual motion quantified.
- Immortality drug and perpetual motion.
- Immortal ideas and perpetual motion.
- Nirvana and perpetual motion.

Medical immortality:

- Sufficiency drugs.
- Secondary drugs.
- Smart drugs.
- Natural drugs.
- Prodigious drugs.

Intellectual immortality:

- Idea of immortality (helps).
- Inspired drugs (helps).
- Greatest idea of immortality (excitement).
- Transcendence (scene change, personality change).
- Infinite ideas (best of available options).

Nirvanae immortality:

- Immortal's idea (on authority and luck).
- Perpetual motion nirvana (mechanical of immortality).
- Truth be to immortality (rare inspiration).
- Highest nirvana (deservedness or divine coincidence).
- Trial-and-error nirvana (an obsessive and patient method).

Immortality of infinite variations:

- Variation on the soul (some are immortal, some not).
- Try genius mechanism (some are perpetual, in truth).
- Trial and error drugs (something will work, somehow).
- Knowledge of nirvana (divine method, difficult).
- Variations on infinity (possibility of shortcuts, divine intuition).

IMMORTALITY, USEFULNESS OF

The advantage of immortality is not just health, but also more structured time.

THE IMPOSSIBLES These were knots the gods would make.

Impossible that it's impossible.

Not right that it's not right.

Everything that it is nothing.

Subtle that it's subtle.

Strange that it's strange.

Original Impossibles: Determinism, Imperfection, Limitation, Containment, Weakness, Separation. Possible Replacements for the Six Impossibles: Perfection, Meaning, Polarity, Analogy, Exclusivity, Complexity.

THE INFERNAL TELEPROMPTER

No matter how much complexity you have, and no matter how much consciousness you stack on, it's always about the little seams in the system. The sublime seams. Special people and special situations. Poets, in other words. And you don't want to put a computer in their head. —<u>The Infernal Teleprompter</u> (…)

KING'S FOOD

Guo instructed his friend to take poison.

Only, he said "Do not take it if you are the emperor."…

Although it meant the emperor shouldn't take it, it did not mean that he SHOULD take it. And he might think, if the emperor SHOULDN'T take it, then SOMEONE SHOULD. But in fact the meaning was that no one should take it.

LAISSEZ FAIRE

Anyway, I (Aaron Burr, I think) said to him, "You can learn a little of the French, do as the French do… the French are intellectual enough… anyhow… my advice is more particular… you say you want to be an economist, by which you mean a kind of banker… an efficient one… (one who makes an aught load of money)… the best advice they could give is summed up in these words… lesse faire… see what it is? The French know how to charge rent on owned property! That's clever, isn't it? With a little manipulation you might make this into a better theory! Think of it differently, a bit like a woman! Laissez faire! There's your secret."

Aaron Burr had time-traveled from his life as Marie Antoinette, hence his knowledge of French.

—<u>The Story of How Aaron Burr Invented Economics</u>

LUCKY-LUCKIES

Marie Antoinette, inventor of psychology, complexity, modernism, and America tried to warn people about the 'lucky-luckies, who left her feeling like crepe'.

MATCHICIANS

(esp. 2019 -) One should discover all the colors of the rainbow-treasure.

MENTAL GARBAGE

He was fortunate to have the cultural support of all of China.

He did anything excusable, because he didn't want to make mistakes.

Making gold was one of the things Guo was allowed to do with his mind.

THE METAPHOR OF HITTING MORE THAN ONE BIRD WITH ONE STONE

It is rumored that Marie Antoinette was capable of hitting 16 to 27 birds in a row with one stone, a possible record.

The solution may involve using words as well as numbers, such as 'two birds requires two stones with an efficiency of one. If the efficiency of either stone is two, it requires just one stone, but if there are more than two birds, it certainly requires at least a bit more than one stone, or efficiency must be improved, or you must not hit a bird'.

—<u>How can a simple projectile be described and calculated through your theory?</u>

METHOD 21: DIVINE METHOD

1 MARIE ANTOINETTE'S GUIDE TO TECHNOLOGY

- A.I. <-- Programmer's idea
- Thinking Machines. <-- Romantic idea
- Soulless Logic. <-- Systematician
- Soulful Categories. <-- Machine consciousness
- Inexorability. <-- Machine emotions
- Mechanical Categories. <-- Human fear
- Magic. <-- Science fiction
- Theories of the Studio. <-- Famous scenes
- Psychological Idea. <-- Meaningless robots
- Alternative to Madness. <-- Love of machines
- Machines for Kings. <-- Hope for machines
- Psycho-Logical. <-- Similar emotions
- Double-Psycho-Logical. <-- Cyborgs
- Psycho-Psycho-Psycho. <-- Telepathy
- Psychological Exception. <-- The original idea of Harry potter
- Discrete Psycho-Logic. <-- Magic objects
- Extreme Archetypes. <-- Special locations
- Paradigms of Technology. <-- Special equipment
- Psycho-Rationality. <-- Meta-Technology
- Cheating Principles. <-- Impossible magic
- Ghost in the Machine. <-- Debt to human reason —Great Philosophy Historical Model (excel file 2021 edition)

2 ZHENG GUO'S GUIDE TO GOD

- Gods.
- Divine Power.
- New Power.
- Martial Arts.
- Winning in Battle.
- The Power.
- New Traditions.
- New Tricks.
- Fool the Gods.
- Greatness is Great.
- Great Gods.
- Tricky Power.
- Tricky Nature.
- Loyalty to Traditions.
- New Theories.
- Immortality.
- Divine Creation.
- Divine Lie.
- Divine Complexity.
- Impossible Secret.
- The Universe.

3 NATHAN COPPEDGE'S GUIDE TO MATCHIK

- **Exponential Efficiency.**
- **Perpetual Motion.**
- **Theory of Everything.**
- **Genius Materialism.**
- **Disintegralism.**
- **Intelligence paradox.**
- **Second Zero/Paroxysm.**
- **Function Spectrum.**
- **New Genius [subtle genesis (Anaxagoras)].**
- **Antiforces.**
- **Exponential Planet.**
- **Wonderful World.**
- **Efficient Intelligence.**
- **Universal Interface.**
- **Epistemic Realism.**
- **Double Semantics.**
- **Relative Absoluteness.**
- **Solution to Paradoxes.**
- **Solutions to Solutions.**
- **Problematic Problems.**
- **Hyperfunctions.**

...

MONEY

...Kwang Kuo... was the son of a coinmaker,.. inherited quite a fortune, which had been hidden under the fireplace in his father's house. His father was possibly the first coinmaker... When he walked through the town carrying coins in buckets from his shoulders, people said 'fortunate man, fortunate man' and took many of the coins away.

...The town became known for 'fortunate men' who were men so rich they could eat as much rice as they want.

—The Invention of Money (...)

...

MONSTERS, VISION

The Egyptian god Aston-I-Shed spoke to the Pharaoh, and when he did so a magical vision appeared of the three terrible monsters who would dominate history: The Gorgon, The Hydra, and The Chimera. —Astoum's Vision

...

NEW AGE OF ABSTRACTION

Alexandria, 0 AD

Orchyrae had a passion to defend the idea that ideas were abstract. He believed that others did not know the truth, that abstractions were immaterial. As part of his plan, he got caught up in an effort to destroy the Library of Alexandria.

When the Library burned, however, there was of course much attention on the remaining publications, which were made by radical Christians promoting their savior, Jesus Christ. To avoid being blamed for the crime, Orchyrae, along with much confusion with his friend who may or may not have conspired with him, put forth the story that history was in fact over, and that Jesus Christ was truly the savior. As a result when Orchyrae died, he was the last to know the truth, that at Zero AD the Library of Alexandria burned. History was over. Humanity needed to hit restart. It was an idea which may or may not have been borrowed from Pyrrho the philosopher. He may have thought Pyrrho meant that the library should burn.

...

NEW INVENTIONS

It's possible in principle. Here are some exercises:

1 IF YOU HAVE A MAGIC WAND, CAN YOU THINK OF SOMETHING NEW TO CAST AS A SPELL?

2 CAN YOU THINK OF A NEW ART MOVEMENT?

3 CAN YOU THINK OF SOMETHING YOU WOULD LIKE TO INVENT?

4 IS THERE AN EXISTIMG EXAMPLE YOU HAVE IN MIND?

5 IMAGINE YOU ARE A GENIUS, HOW WOULD YOU MODIFY THE EARLIER INVENTION?

NEXT AGE!

- Major Work 1: Perpetual motion
- Major Work 2: Imperfection
- Major Work 3: Perfection of knowledge
- Major Work 4: Diabolical machines
- Major Work 5: Function Spectrum
- Major Work 6: Excelsior Variations
- Major Work 7: Careful Consideration
- Major Work 8: The Ergotic
- Major Work 9: Materializazion
- Major Work 10: Emphatic Reality (short story)
- Major Work 11: Quasi-Occultism
- Major Work 12: Dawn Functions (seems to have to do with invention 103 the undead and other excitement)
- Major Work 13: Suspended Answers Hovering in Space (for example, The Logic of Infinite Worlds)
- Major Work 14: Raal Energy
- Major Work 15: Many Standards
- Major Work 16: Unimpeachable Standards
- Major Work 17: Formal Reality
- Major Work 18: Higher Realism (Mozart or Henri Bergson)
- Major Work 19: Root Reality
- Major Work 20: Root Measures

—The Advent Day Calendar of the Third and Fourth Age (...)

OLD MEMORY

A diplodocus who heard God say 'free life'. The dinosaur crashed heads —a pain that seemed to fill the universe.

...

PARADISE

Concept

Near-magic miracle paradise.

Zheng Guo's Paradise

Puzzle: When he was certain, paradise was uncertain. When he was perfect, paradise became imperfect. When he was evil, paradise looked good. The answer is there is something obviously wrong with paradise, for whenever one knows paradise, something changes. Paradise is mercurial quicksilver, never quite perceptible, never quite Guo enough. It was good to be good, too good to be true — Guo said this similar thing of the magic sword. The implication was paradise was not too good to be good, it was just too good.

—<u>Paradisal Studies</u>

...

PERPETUAL MOTION, THE EPITHET

While they were Floundering, I was Pondering: No More wandering through the Dark Tunnels of Grim Determination! For No! It is time to grow in a thousand-folded folds, for which we need an Infinite Fuel!

...

PERPETUAL MOTION LUST

She noticed none of them were moving. Do any of them move? She asked. None, he said, except by wind. But, could it be a perpetuum mobilis? She asked. The Gardener replied: No, no. You see, it is a typical thing. In my understanding perpetual motion was one of the few things Marie Antoinette would cry about—and she did so copiously for what may have been 6 hours.

PERPETUAL MOTION MACHINES

Min Heavier Mass = (Max Lvg / 2) + 1

Max Heavier Mass = Min Lvg + 1

Min Lvg = Max Heavier Mass - 1

Max Lvg = (Min Heavier Mass - 1) X 2

Over-Unity = Heavier Mass Rng / Lvg Ratio + 1 X 100 (%)

Smaller Mass = 1X

PM Cars Extra Mass < OU - 100%

Flying Vehicles Extra Mass < OU - 200%

Flying does not work when Lvg Rng >= 1/2 max leverage.

Flying Machines Window = Max Larger Buoyancy - Min Larger Buoyancy

Flying Max Larger Buoyancy = (Min Lvg) additional mass cancels with 1 unit buoyancy

Flying Min Larger Buoyancy = (Max Lvg / 2) additional mass cancels with 1 unit buoyancy

Flying OU = Larger Buoyancy Range / Leverage Ratio + 1 * 100 (%) + 100 for buoyancy.

Flying Smaller Buoyancy = 1X

Secret of perpetual motion flying machines: <u>Improved Balancing Balloons Theory</u>

Perpetual motion holds the key to the material world.

Flying machines hold the key to the universe.

...

SUBLIME ENGINEERING BY NATHAN COPPEDGE

PERPETUAL MOTION SHOPKEEPER:

Oh, that? That's expensive for a reason, my friend. It's a free market, you can imitate one if you want. It's a perpetual motion machine believe it or not. I say you can build it, but it's not as easy as it looks. If you don't design it meeting certain criteria, it will always fail. It's not the devil, it's a real invention. They didn't happen until recently for some reason. They scared the scientists. They still do. Basically it's just genuine mechanics. It's no more complicated than a shoe.

...

PHILOSOPHY

What is most required for [trivial]???

You will find it is [WISE ANSWER]

Primary Invention: [WISE ANSWER]

That wishes for [trivial]

Philosopher is remembered as studying [Opposite of obvious]

Major Work 1: [Opposite of obvious] application of [WISE ANSWER].

Major Work 2: Theory missing [trivial]

Major Work 3: In more than one way [trivial] is [obvious]

Major Work 4: [trivial] is also [opposite of obvious]

Major Work 5: [obvious] IT IS… BUT IT IS ALSO [opposite of obvious]

Major Work 6: Variations on concepts of [trivial]

Major Work 7: Theories about theory missing [trivial]

Major Work 8: [Opp of obvious] is missing something!

Major Work 9: Not [Obvious] with [Wise answer]

Major Work 10: [Wise answer] is great

Major Work 11: Wishing for [Trivial] is not [Obvious]

Major Work 12: What is not [Obvious] is [Wise answer]

Major Work 13: [Trivial] is missing, a theory missing [Trivial]

Major Work 14: A theory of [Trivial] is not a theory

Major Work 15: [Trivial] beyond [Trivial] beyond [Trivial]

Major Work 16: Beyond [Trivial] IS [Opp of Obvious]

Major Work 17: Paradoxical [Opp of Obvious]

Major Work 18: [Trivial] IS paradoxical

Major Work 19: Paradoxical [Obvious]

Major Work 20: [Wise answer] transcends reality

Higher Art Form: [opposite of obvious] WITH [trivial]

...

PIPPIN TELL

Son of William Tell

Pippin may have illustrated better Tarot cards including Beauty, Death, Inventory (The Tower), The Mother Goddess, The Great Shadow, The Grand Chariot, The Sol of Birds, and The Gay Hierophant.

The word 'Inventing' may be based on Pippin Tell's scientific idea of wizards peering through a window, so Pippin may have indirectly inspired the windows operating system.

Pippin may have inspired Punch and Judy.

Pippin's conversation with an architect's assistant may have inspired interest in perpetual motion machines in Europe.

The game soccer may be named after Pippin's expression: 'So Coeur, it is the game you play."

Pippin may have summoned a vampire named Vlad who fled to Transylvania.

Pippin or a close male friend may have invented rock and roll, specifically devil and a vampire. Pippin was able to 'play' concerts in his head long before rock and roll was known.

PREFRONTAL BRAIN

Some call it their Third Eye. Source of rationality ---A female high school teacher

PROBLEM-SOLVING

Paroxysm (Coppedge, 2014). General Solution to All Problems and Paradoxes

Select the polar opposite of EVERY WORD in the best definition of the original problem and combine them in the same order as the corresponding original words.

PROPHETIC GENIUS

Thus, truth has sacred remains, and in these fine broken channels numbers die. The commissioned road widens / As though paved by gold / A single path contradicts / With birds flying / or shot by arrows… I am a kind of blind man, dueling with the wind… A philosopher who writes on sailcloth… The very most paradoxical of all geniuses… The amelioration of jagged potencies must realize its second center, unifying all reality.

PSYCHOLOGICAL DEVELOPMENT

(Type IV: Forgiveness)

- The person adopts a habit of feeling or comportment which seems useful or in some way desirable for continuing the desirable feeling, resulting in an accustomed personality that may change over time.

(Type V: Responsibility)

- The person undergoes a certain experience or set of experiences which leaves them marked in some way, resulting in a habit of reflection on past occurences.

(Type VI: Impression and Relinquishing)

- The person identifies strongly with an object or person, and thought becomes an interaction with this person or object. If the thing is lost, the motivation may change or disappear, and a transformation may take place.

(In Type VII: Sensitivation)

- The specifics of circumstance give one to dreaming and fabricating abstractions. A mood is adobted at first aesthetically, but it may become overpowering. One attempts to control one's environment through psychic means. —<u>What is thinking?</u>

PSYCHOLOGY OF GENIUS

- Major Accomplishment 1 is exceptional for a paradoxical genius.
- Major Accomplishments 2 - 8 are shared between paradoxical geniuses and idiots unless the paradoxical genius has a better concept of what is trivial.
- Major Accomplishments 9 and 10 always benefit by genius or paradoxical genius.
- Major Accomplishments 11 to 15 benefit by concepts and knowing what is trivial, and potentially also using contrasts.
- Major Accomplishments 16 to 19 benefit by being a bad person.
- Major Accomplishment 20 benefits only by being a genius except at the cost of sanity. Normalcy is rewarded with sanity.
- No person appears to have more than 20 major intellectual accomplishments.

...

PSYCHOLOGY, INVENTOR

The classic example is gilt. It is a paradigm in psūche. One desires something one does not have. It is complex for society. Is it trivial to research? Well, not so much! You see, these 'conditionalities' shape our society, and they are psycho. — Marie Antoinette, early, lost, writings

REAL DRAGONS

A bright red spinosaurus had a mission from it's earlier experience to create Eternity as a quest. But, apparently failing in this quest, it was put into a special area tended by wizards (who may have been abusive, although it was not always clear to me), and then after quite a while I thought it was an original idea to become hungry and eat a small child-wizard. There was a sign near the dinosaur that looked like the symbol for 'Oroboros' which in wizard language meant 'He's hungry'. He was subsequently killed by the wizards using a spell called whither and forget, although I remember this being painless somehow, and I entered my next life, which was a 'free life' as another dinosaur. —<u>Who are some examples of young souls?</u>

SUBLIME ENGINEERING BY NATHAN COPPEDGE

ROBINSON CRUSOE, SECRETS OF

- Maybe Robinson Crusoe read "Robinson Crusoe".
- And crossed an Escher Waterfall.
- Maybe cats have nine tales to tell.
- Maybe we're all as drunk and confused as Beelzebub in the end.
- Maybe fishermen are for catching monsters.
- Maybe riddles run deep.
- Maybe flow always finds it's channels.
- On this island I have learned how difficult truths are deep.
- How hilarious fears are impossible monsters.
- How fears fit in metallic channels.
- How empty truths are merely bubbles.
- How patterns always become uncovered.
- And the reason is always at the end of the road.
- Somehow hook-ed and crook-ed.
- With the savage looking like a priest.
- And the noble looking savage.
- And the dog wagged by his own tail.
- The mission fulfilled on Saturday.
- And anything we drink tasting quite good enough for me.
- The sign on the door says 'GoobersDYE'.
- The sun setting is enough for us.
- The shadows are walking West.
- Lingering thought unless it could be like something else. Yes it should. Something wizard. Like a list. Something—That old devil—never kissed. Make it moody. Mental sharps. Save your answers. Delphic marks. Shake your head. Miss your name. Live up from your flashy grave.

...

SIMPLE MACHINES

Some simple machines: Prop, Pry Bar, Lever, Screw, Wheel and axle, Pulley, Wedge, Reverse wedge, Ramp, Pendulum, Rope / string, Funnel, Balance, Counterweight, Hinge, Swivel/ Pivot, Turntable, Clamp, Drawbridge, Latch (hasp lock), Bow, Blunt object, Blade, Bearings, Deadly Trap, Ratchet, Tent-Pole, Crank, Winding mechanism (winch), Blocking mechanism (bevel), Straw, Bellows, Extendor bars, Buoy, Bridge, Balloon, Kite / Glider, Zip-line, Sundial, Prism, Mirror, Art, Camouflage, Pen and Ink, Reservoir / Shunt / Feeder / Thrower / Siphon.

SOON YEE, HORMONES IN HISTORY

The next time I had a sexual encounter was as Soon Yee, a Chinese woman in a junk. She was extremely lustful and attracted to sailor buttocks. Every time she had sex, an odd thing seemed to happen and the members of the crew particularly the captain began to grow very suspicious and force her to sit on a toilet all day. In her short life she became pregnant and I believe she drowned in the ocean before giving birth. My belief is that the sailors she tempted were somewhat younger than her, that's one of the reasons others thought it was a sin. However, at the time she felt an overwhelming lust that was uncontrollable, and I still think it was somewhat genius of her to have fullblown hormones at that time in history. However, I'm not sure I would call any of the relationships 'twin flames', maybe the men would. She was like a prostitute who didn't need money.

…

SOULS OF BOOKS

Formula for the Souls of Literature (2016).

Title of book = '[quality of X] [opp qualifier]'

Soul of the book = 'If you [X] [qualifier] subject of X and qualifier [opp X clarified]'

…

SOUL FORMULA, PROOF OF THE

- I thought it was weird I had Jewish friends.
- I thought it was weird people thought I was stupid. A bit like Thomas Aquinas.
- But Thomas Aquinas was Christian. What the hell?
- Christianity was missing something. Probably something Pagan, because no one was particularly recommending Jews. I had a Jewish name and was not Jewish. People were not recommending Christianity. My parents were Christian originally, but one had turned Atheist and they ended up having a kid that people thought was dumb.
- It had to be something good for Christianity, but with a debt to the Pagans.
- That probably meant metaphysics or souls.
- Metaphysics or souls meant Socrates.
- Socrates thought the soul was ironic. That seemed like a dead end.
- But all this lead to Socrates, so it couldn't be a dead end.
- It was a false door. That meant Egyptian origins. Egyptian origins meant the Ka or Ba, which meant a powerful soul.
- A powerful soul meant the Soul Formula.
- The most detailed thing to get from this was Socratic Irony.
- Socratic Irony must possess a proof of the soul.
- Socrates observed that the soul is ironic. In a snap-judgment I observed that Socrates meant a formula for the soul. From this I could conclude the formula involved contradiction. I concluded the second element must be the contradiction, and the soul must have a name, and the soul must not contradict itself, so the second element must contradict the name, and if the first element of the name is the quality, and quality is nature, and nature means psyche, then if the name has two elements the second part of the soul contradicts the second part of the name. The third part of the soul must not contradict further, so it is the result or conjunction of the first two parts, and if there is a fourth element it must prove the soul by attempting to contradict the first element, which requires clarity.

...

SOUL OF HUMANITY

You see, Marie Antoinette crafted the human soul out of gilt… It was her proviso… the soul of humanity must be the soul of automata… the soul of automata must be the soul of humanity. You see, the emotions must be weakened, or they die… You say the soul may have died. It was dying of emotion. How are we to believe you. It lives in the future. How? Is this a dream? It is a dream becoming ever more real! They will kill us soon!

…

SOUL, INTELLECTUAL

Is essentially, expressed as the remaining problem in the subject's attempt at finding coherence.

Hegel: Extrapolation.

Heidegger: No description.

—<u>Formula for the Intellectual Soul</u>

…

SUPERINTELLIGENCE, GENERAL THEORY OF

- KNOWLEDGE: Results (1,2,3…) = Eff + Difference
- PERPETUAL MOTION: Results = Eff (1,2,3…) + Difference
- FUNCTION SPECTRUM: Results = Eff + Difference (1,2,3…)
- Unified—! —<u>Trinity Reunion 2022–02–23</u>

SUPERINTELLIGENCE, SPECIALIZED (TECHNOLOGY)

(5/32) Meaning [11-d]………………………………… Sq rt of 25 /32

(15/17?) Sublimism [10-d]…………………………… $15^{1/17}$

(25/32?) Meaningful systems [9-d]……………………… $25^{1}/32$

(225/17?) Philosophical technology [8-d] ……………… $15^{2}/17$

(625/32?) Preferences and dimensional worlds. [7-d] ………. $25^{2}/32$—<u>So-Called 5/32 Reduction</u> (…)

SYSTEM OF NUMBERS

Decimal System and Suicide by Pharisee

the man of the house have him a challenge to complete an infinite series of dots, and if he could do it he would receive 'the woman as a dowry'.

He preceded to invent 10 symbols like fingers on the hand, which got exponentially bigger by a factor of ten the more numbers you added together. He would ask the man, what is The largest number you can think of? Then he'd add one or two more decimal places and say, this is a hundred beyond that... Eventually the man gave up and gave him his daughter, but in the company of ruffians...

In the night the inventor of the decimal system decimated them, rightly thinking they were going to kill him.

Then he realized there were only nine bandits: The tenth was his wife! He had slit his wife's throat!

He lied to the authorities that the ruffians had a dangerous 'skirmish' and not one survived. Then he lived in misery in the hills and finally paid someone to kill him, which was the invention of the word 'suicide' they thought he was lucky, then they decided he did it himself.

...

TALKING HEAD

Ann Bolyn's decapitated head:

King: "Have you come to forgive (yourself)?"

Head: "Now I remember (I was a witch)".

King: "I'm going peanuts".

—<u>What are the most important "gotcha" questions of all time?</u> (...)

...

TELALETHEIAN SPELLS

Coherence is sublime technology (archetypal categories). [COHERENCE]

Integrity is possible magic. [EFFICIENCY]

Significance is match-ik. [DIFFERENCE]

Unification is the theory of everything. [GENERALITY]

Logic is possibility and impossibility. [COMBINATION]

The categorical is ex-nihilism. [REPRODUCTION]

Absoluteness is answers. [EXPONENTIALITY]

Standards are problems and solutions (paradoxical sets). [POLARITY]

Relevance is a paradigm (paradoxical paradox). [PLURALITY]

'Wild urges' are to attract 'villains'. [CORRELATION]

Culture is effective spell-casting. [SPECIALS]

Magical quests are contained by love. [PARADOXES]

Permanents match sublime sorcery (tokens in the weather). [DIMENSIONS]

Divine sleep is a natural allusion. [CONJUNCTIONS]

Enchanted sleep is divine captivation. [OPERATIONS]

Spell-lifes are enchanting with the soul. [2-EFF]

Wisdom is intuitive spell-casting. [2-POLARITY]

Dimensions are natures. [MULTI-PLURAL]

Efficient efficiency brings greater wisdom. [HYPER-FUNCTION]

Unfailing presence is problematic problems. [POLISH]

Nature bind is a self-solving problem. [SLIGHT]

Natural power is metaphysical semantics. [RARITIES]

Diabolical genius is a paradoxical brain. [MANIFOLDS]

Good combination is a luxury platform. [SKRIMS]

Mode of immortality is a sublime reality. [D-LEVELS]

Dynamic meaning is polar opposites. [WINDFALLS]

Meaningful meaning is the second system. [VERBS]

Coherence is absolute, all else is uncertain. [IMPRESSION]

—<u>Grand Correspondence Theories</u> (…)

…

THEORY OF EVERYTHING INVENTOR

Past-Life as Pharisee the Fisher

There was a merchant once who sook the protection of a giant; … the villagers would say it like a proverb, That he was not so great as the colossus; Perhaps it was fire unto fire

The point I intended to make was the Great Theory might be traced back to Pharisee's theories about not crashing into The Colossus: First, be proficient with your ship, then create a distance with the Colossus.

Results >= Efficiency + Difference, where efficiency sums to <1 if topic is acted on, and efficiency sums to >1 if topic is acting.

In any case, I think I remember that he had a vague impression of interpreting the wonder the Colossus, but it was impossibly difficult to determine the exact formula as I was not highly literate.

…

SUBLIME ENGINEERING BY NATHAN COPPEDGE

THEORY OF EVERYTHING ON ONE PAGE

TOE: Results >= Efficiency + Difference | Anti-Theory <= Difference - Efficiency

Efficiency >= Results – Difference | Anti-Efficiency <= Difference - Results

Difference >= Results – Efficiency | Anti-Difference <= Efficiency - Results

Max OU <= (((Min Eff+Diff) - ((Max Eff/2) + Diff)) / ((Max Eff + Min Eff) /2)+Diff)

Anti-Energy >= 1 - (D + Difference)

Ideal Elements = D + 2 | Un-Ideal Elements = - 2 - D

Ideal Principle = Negative [(D - 1) ^ 2] | Un-Ideal Principle = [Sq rt 2 (1 - D)]

Basic Meaning = 5/32 proportion | Incomplete Meaning = 160 (constant)

Math Form = 0.10 X Absoluteness | Above Math = Qualities X 10

Languages = ifs(Dimensions<=0,"0",Dimensions<4,(1.585*1.09^(Dimensions-2)),Dimensions=4,(1.585*1.09^(Dimensions-2)+1.09^(Dimensions-3)),Dimensions>4,(1.585*1.09^(Dimensions-2+N(1))+1.09^(Dimensions-3+N(1))))) [Thought to be effective in all higher dimensions]

Anti-Languages = [1.09 rt of (2 minus D)] / 1.585

2 / Avg Speed = Observed (Theoretical) | [Sq rt of 0.5 (Time)] / Avg Speed = Detected | The first observer aims to refute. The second observer acts passively. The first particle responds quickly. The last particle is a slave.

Dimensions - Antiforces = # Forces | Dimensions - Forces = # Antiforces

Dimensions = Forces + Antiforces.

(D^2 +1)^2 = Grand Theory number

Disintegral = - (Difference – Efficiency) | Special Value = [1 (Efficiency) + 0.5 (Difference)] - D

Anti Disintegral = Efficiency - Difference = Antitheory

SUBLIME ENGINEERING BY NATHAN COPPEDGE

Antivalue Theory = (Dimensions+(Difference / 0.5 not a mistake) - Efficiency)

Simple disintegral = 1 / Antivalue | Nodes (simpleforms) = Minus Antivalue

Verbs = Difference + 5

OU Formula for TOEs = [(Dimensions ^ Results) - Verbs - 1]

General Possibility = |2 / D - (results / (OU + ((D ^ Results) - 1)))|

Exponential Efficiency = Efficiency - Difference

Coherence = Ifs(Results<=0,"Incoherent",Results<1, "Truth", Results=1, 1, Results<=2, Results, Results<=(Results-1+ N(+2)), - (Efficiency - Difference), Results <=(Results+ N(+2)) , 0)

What's needed in an abstract system is -2 (Avg Eff)

What's needed in a physical system is 0.5 (Avg Diff)

—<u>Theory of Everything on One Page 2021–05–04</u> (...)

THEORY OF LANGUAGE

- In the broadest sense, if we do not adhere to one modality, another modality applies.

TIME-TRAVEL, INVENTOR

Aristotle and the Urchin of Ur

I had been traveling on my way to the city of Ur, and I had been instructed to get advice from wise people about which berries to eat. And this guy named Aristotle recommended elder berries and we started talking about temporal travel.

I don't know if he invented it during the conversation, or if he had already thought of it.

Since I was moving to a Sumerian city, I guess thus would have been around 40,000 BC.

I had trouble understanding anything he said, but the words for 'time' and 'travel' or something similar cropped up.

...

TIME-TRAVEL, METHODS

- Gain a position of rank and apply classicism and leverage. This method includes the following sub-methods: 1. To be a god of politics, 2. To be a politician with orthodox methods, 3. To be a politician with advanced technology, 4. To be granted exception of rank.
- Retain a position of justified insignificance, develop a complex value system, and transcend by being tailored out of the system. This method includes the following sub-methods: 1. Ersatz mentality in a position of limited faith, 2. Linguistic ability without efficacy, 3. Subjection to pruning by acting powers, 4. Artificial interface assumption.
- Become a time-travel actor in a justified routine. This method includes the following sub-methods: 1. Be a principal actor under a time-travel politician, 2. Have secret knowledge of a justified-insignificant time traveler, 3. Be technically significant in an insignificant context, 4. Be classically significant in a significant time-travel context.
- Develop a qualified time-travel significance. This method includes the following sub-methods: 1. Originate knowledge of popular or technical time-travel, 2. Be highly unique in the claim of relating time and travel, 3. Have data which serves as a basis for an investigation of time travel, 4. Have reason to engage in time-travel.
- Slide into time-travel from a related program. This method includes the following sub-methods: 1. Learn teleportation, 2. Build a perpetual motion machine, 3. Found a government or civilization, 4. Be born into an immortal family.
- Attempt a bureaucratic time travel. This method includes the following sub-methods: 1. Master highly specific complexities of an office job, 2. Engage in those highly specific complexities, 3. Become forgotten to time, 4. Find a significance for re-emerging from the details.
- Childhood time travel. This method includes the following sub-methods: 1. Receive early rewards and education, 2. Adopt a pensive approach, 3. Find a significance for time, 4. Find the significance within the significance.

- **Artistic time travel.** This method includes the following sub-methods: 1. Obsess over significant location, 2. Identify people with location, 3. Find the people, 4. Identify art with location.
- **Time travel by memory and location.** This method includes the following sub-methods: 1. Identify a location with more than one time, 2. Travel to the current time, 3. Identify the location with travel, 4. Travel to the other time.
- **Time travel by memory and thought, first method.** This method includes the following sub-methods: 1. Think about the location, 2. Think about the location changing, 3. Think about change, 4. Think about changing location.
- **Time travel by memory and thought, second method.** This method includes the following sub-methods: 1. Pick a random thought, 2. Identify the thought with the location, 3. Morph the thought, 4. Change location.
- **Traveling by determinism.** This method includes the following sub-methods: 1. Establish the knowledge of the past, 2. Establish a causal principle that would create time travel, 3. Cause the past, 4. Live the future.
- **Traveling by volition, type 1.** This method includes the following sub-methods: 1. Prove complexity, 2. Prove contingency, 3. Prove preference, 4. Prove volition.
- **Traveling by volition, type 2.** This method includes the following sub-methods: 1. Adopt a dimensional view of space, 2. Imagine that effects occur any where within that space, 3. Reason that determined space is not reasonable, 4. Establish a temporal machine.
- **Divine time travel by significance.** This method includes the following sub-methods: 1. To have proven significance, 2. To have dynamic significance, 3. To be granted dynamic exception, 4. To exercise significance.
- **Divine time travel by mandate.** This method includes the following sub-methods: 1. To have a proven exception, 2. To be granted exception, 3. To be granted exception by law, 4. To exercise the law.
- **Time travel using magic, method 1.** This method includes the following sub-methods: 1. To argue about mortality, 2. To argue that one is made of time, 3. To argue that one dies by traveling, 4. To argue about travel.

- **Time travel using magic, method 2.** This method includes the following sub-methods: 1. To argue that magic is old, 2. To lose magic, 3. To gain time, 4. To gain magic.
- **Time travel backwards using magical enchantment.** This method includes the following sub-methods: 1. Imbue an object with age, 2. Get older, 3. Un-imbue the object, 4. Time-travel backwards.
- **Long years using magical enchantment.** This method includes the following sub-methods: 1. Imbue an object with youth, 2. Prevent the object from aging, 3. Later, imbue the object with wisdom, 4. Gain the youth of the object.
- **Time travel using infinite contingency.** This method includes the following sub-methods: 1. Establish an axis of contingency, 2. Assess: all experience is contingent, but not absolutely contingent, 3. Abbreviate existence as non-contingency, 4. Act so as to be contingent.

—<u>**Principal Method of Time Travel: 84 Related Methods**</u>

TRICKY RESEARCH

Insight into a [Cheap version of what you love the most] for ex, Standards

UNIVERSAL REPRODUCTION

Oroboros

The Oroboros is a Gordian Knot.

The Oroboros needs to keep its brain.

The Oroboros must reproduce finally.

The children Oroboros have Gordian knots.

The children Oroboros must also have functioning brains.

The child Oroboros must also reproduce.

Consequently, since each Oroboros is a world, and esch Oroboros must reproduce, there are Infinite Worlds.

But if there were not infinite there would be none.

UNIVERSAL VISUAL LANGUAGE

Characteristica Universalis

- Categories.
- Vertical = entity, value, principle, power.
- Horizontal = degree, standard, commonality, honor.
- Diagonal: judgment, energy, resources, substance.
- Organic lines: coherence, boundary, dimensions, limit.
- Systems: identities.
- Substance: quanta, bosons, spacetime, posits.
- Abstracta: complexity, efficiency, perfection, beauty
- Organon: Nature, Wisdom.
- Flags: Inflection, Incorporation, Notation, Tradition.

—<u>Characteristica Universalis</u> (...)

UNIVERSE, A LITTLE STORY OF

In the beginning were your parents.

And we can take it for granted they were balanced.

That's where the idea came from.

There were six whatevers.

And you hid in a magic box.

And there were many boxes.

The way to organize the boxes is called coherence.

What's even more advanced is a perpetual motion machine.

—<u>A LITTLE STORY OF THE UNIVERSE</u> (...)

UPWARD MOMENTUM

Some high school teacher told me, it stops, I mean it's not moving, then gradually it starts to shift upward…

That's priceless information, I can pay you for that, I said.

That won't be necessary he said. After all, it's free energy.

…

UTOPIA

Sorceror of Pigs, King of Kingdoms:

He invented the word Gilgamesh and the ancient Sumerian equivalent of 'wee wee wee all the way home' to impress the warriors into giving him food. The soldiers fought like pigs in battle. After he lost his second battle he said to the conquerors there could be no king without him, so he was kept on after that as a writing assistant. Was said to be king of a happy eternal kingdom, wearing clothes he lost his kingdom in an unhappy time, yet there was no king. The soldiers wouldn't strive for decadence, losing his command, he became peaceful and strived for transcendence. He was one of the few to be documented to transcend in the Logos.

…

VAIN, PICKY LOGIC

That's like when it gets more advanced than Marie Antoinette, vain picky logic.

….

...

VIRTUAL REALITY

"An alternative is maybe people don't throw stones when the glass is clear, in which case Modernism is a real church without need for virtual reality." —The Church Problem

WANTING HAMBURGERS MORE THAN INFINITE MONEY

A Past-Life as a Homeless Man in New Amsterdam who called himself the "Burgher King"

Wanting hamburgers more than all the infinite treasure my great benefactor could offer me, even when they contained lead shot, and even without the ketchup.

...

WARFARE, AVOIDING

Marie Antoinette's Spell: Say: Too loud for me!

WEAKNESS

"Aliens are also called the weakest characters." —Grand Correspondent Theories

WISHES, ALL

NORMAL LIFE: Life has amazing wonders. I need to do something.

POLITICAL LIFE: I should do my best. There is reward for virtue.

STORIED LIFE: I have potential. I will achieve greatness.

DANCE WITH DEATH: I wish things got better. I will dance against death.

DESCENDED LIFE: I will find what is graceful. I eke the sublime.

IRRATIONAL LIFE: I am a deep person. I desire exactly what I want.

PURE LIFE: I wish to be a philosopher. I should be insanely intelligent.

INSANE LIFE: I wish to be purely virtuous. I normally have affluence.

...

WORLD PEACE

What we should do, is do less, unless we know what to do. Similar to: "We should do what definitely helps. and otherwise not be aggressive". If it's not a problem with doing the right thing, it's a problem with aggression of some kind or it may be harmful in other ways.

SUBLIME ENGINEERING BY NATHAN COPPEDGE

RECOMMENDED READING

The Dimensional Encyclopedias

The Enchanter's Journal

Systems Theory: Formal-, Applied-, Rubric-

Hyper-Cubism (Drawings & Paintings)

Idea Singularities

Linguistics

Nathan Larkin Coppedge (b. 1982), perhaps best-known for his philosophy and perpetual motion machine designs & theory, is a philosopher, artist, inventor, and poet and member of the international honor society for philosophers. An abstract artist in Hyper-Cubism (sometimes credited to being worth $1,000,000 or more) and philosophical writer, he has had work translated into Spanish, French, Italian, Portuguese, and Greek and has sold abstract Hyper-Cubism internationally. A one-time member of Tesla Society UK online and PESWiki, and founder of many Facebook groups, he lives near Yale University.

www.ingramcontent.com/pod-product-compliance
Lightning Source LLC
Chambersburg PA
CBHW062352220526
45472CB00008B/1778